TEACHING SCIENCE

Science has never been more important, yet science education faces serious challenges. At present, science education research only sees half the picture, focusing on how students learn and their changing conceptions. Both teaching practice and what is taught, science knowledge itself, are missing. This book offers new, interdisciplinary ways of thinking about science teaching that foreground the forms taken by science knowledge and the language, imagery and gesture through which they are expressed.

This book brings together leading international scholars from Systemic Functional Linguistics, a long-established approach to language, and Legitimation Code Theory, a rapidly growing sociological approach to knowledge practices. It explores how to bring knowledge, language and pedagogy back into the picture of science education but also offers radical innovations that will shape future research.

Part I sets out new ways of understanding the role of knowledge in integrating mathematics into science, teaching scientific explanations and using multimedia resources such as animations. Part II provides new concepts for showing the role of language in complex scientific explanations, in how scientific taxonomies are built, and in combining with mathematics and images to create science knowledge. Part III draws on the approaches to explore how more students can access scientific knowledge, how to teach professional reasoning, the role of body language in science teaching, and making mathematics understandable to all learners.

Teaching Science offers major leaps forward in understanding knowledge, language and pedagogy that will shape the research agenda far beyond science education.

Karl Maton is the creator and architect of Legitimation Code Theory.

J. R. Martin is a world-leading authority in Systemic Functional Linguistics.

Y. J. Doran is a leading young scholar combining both frameworks in research.

All three are members of the LCT Centre for Knowledge-Building.

Legitimation Code Theory
Knowledge-building in research and practice
Series editor: Karl Maton
LCT Centre for Knowledge-Building

This series focuses on Legitimation Code Theory or 'LCT', a cutting-edge approach adopted by scholars and educators to understand and improve their practice. LCT reveals the otherwise hidden principles embodied by knowledge practices, their different forms and their effects. By making these 'legitimation codes' visible to be learned or changed, LCT work makes a real difference, from supporting social justice in education to improving design processes. Books in this series explore topics across the institutional and disciplinary maps of education, as well as other social fields, such as politics and law.

Accessing Academic Discourse
Systemic Functional Linguistics and Legitimation Code Theory
Edited by J. R. Martin, Karl Maton and Y. J. Doran

Building Knowledge in Higher Education
Enhancing Teaching and Learning with Legitimation Code Theory
Edited by Chris Winberg, Sioux McKenna and Kirstin Wilmot

Turning Access into Success
Improving University Education with Legitimation Code Theory
Sherran Clarence

Teaching Science
Knowledge, Language, Pedagogy
Edited by Karl Maton, J. R. Martin and Y. J. Doran

Decolonizing Knowledge and Knowers
Struggles for University Transformation in South Africa
Edited by Mlamuli Nkosingphile Hlatshwayo, Aslam Fataar, Hanelie Adendorff, Paul Maluleka and Margaret A. L. Blackie

Enhancing Science Education
Exploring Knowledge Practices with Legitimation Code Theory
Edited by Margaret A. L. Blackie, Hanelie Adendorff and Marnel Mouton

Legitimation Code Theory
A Primer
Karl Maton

For a full list of titles in this series, please visit:
https://www.routledge.com/Legitimation-Code-Theory/book-series/LMCT

TEACHING SCIENCE

Knowledge, Language, Pedagogy

Edited by Karl Maton, J. R. Martin and Y. J. Doran

Routledge
Taylor & Francis Group

LONDON AND NEW YORK

First published 2021
by Routledge
2 Park Square, Milton Park, Abingdon, Oxon OX14 4RN

and by Routledge
52 Vanderbilt Avenue, New York, NY 10017

Routledge is an imprint of the Taylor & Francis Group, an informa business

British Library Cataloguing-in-Publication Data
A catalogue record for this book is available from the British Library

Library of Congress Cataloging-in-Publication Data
A catalog record has been requested for this book

ISBN: 978-0-815-35576-2 (hbk)
ISBN: 978-0-815-35575-5 (pbk)
ISBN: 978-1-351-12928-2 (ebk)

Typeset in Bembo
by SPi Global, India

CONTENTS

FIGURES

TABLES

CONTRIBUTORS

Y. J. Doran is a researcher at the University of Sydney who focuses on language, semiosis, knowledge and education, spanning the interdisciplinary fields of educational linguistics, multimodality, and language and identity. He works primarily on English and Sundanese, and from the perspectives of Systemic Functional Linguistics and Legitimation Code Theory. His book, *The Discourse of Physics: Building knowledge through language, mathematics and image*, was published by Routledge in 2018. He co-edited (with J. R. Martin and Karl Maton) *Accessing Academic Discourse* (2020) for the Routledge LCT series.

Karen Ellery is a Senior Lecturer in the Science Extended Studies Programme of the Centre for Higher Education Research, Teaching and Learning at Rhodes University, South Africa. Her research centres on enabling epistemological access in the sciences in a higher education context, focusing specifically on curriculum structures, pedagogic practices and student learning. Recent journal publications include 'Congruence in knowledge and knower codes: The challenge of enabling learner autonomy in a science foundation course' (*Alternation, 2019*) and 'Legitimation of knowers for access in science' (*Journal of Education*, 2018).

Jing Hao is an Assistant Professor at the Pontificia Universidad Católica de Chile (PUC). Previously she worked as Postdoctoral Research Fellow at PUC and at The Hong Kong Polytechnic University, following her doctorate in linguistics at the University of Sydney. Her research explores knowledge-building through English and Mandarin Chinese and its interaction with other semiotic modes. Her book *Analysing Scientific Discourse From a Systemic Functional Linguistic Perspective* was published by Routledge in 2020.

Susan Hood is an Honorary Associate Professor in the Department of Linguistics at the University of Sydney. Her research draws on systemic functional linguistics in studies of academic discourse that explore disciplinary differences in the expression of knowledge and values, the multimodal cooperation of language and embodied paralanguage in lectures, and storytelling as a research practice. Recent publications include *Appraising Research: Evaluation in academic writing* (2010, Palgrave) and editing *Semiotic Margins: Meaning in multimodalities* (with S. Dreyfus and M. Stenglin, 2012, Continuum). She also co-edited (with Karl Maton and Suellen Shay) *Knowledge-building: Educational studies in Legitimation Code Theory* (2016) for the Routledge LCT series.

Sarah K. Howard is an Associate Professor of Digital Technologies in Education, SMART Infrastructure Facility Education Group Leader, and a member of the Early Start Research Institute at the University of Wollongong. Her research focuses on the use of new methodological approaches to explore teacher change and classroom practice, specifically related to technology adoption and integration in learning. As part of this agenda, she uses Legitimation Code Theory to explore cultural and individual factors of teachers' digital technology use, educational change and learning design, with particular interest in subject-area change and underlying principles of teaching and learning.

J. R. Martin is Professor of Linguistics at the University of Sydney, where he is also Deputy Director of the LCT Centre for Knowledge-Building. Recent publications include a book on teaching academic discourse on-line (*Genre Pedagogy in Higher Education,* Palgrave Macmillan 2016), with Shoshana Dreyfus, Sally Humphrey and Ahmar Mahboob; a book on Youth Justice Conferencing (*Discourse and Diversionary Justice,* Palgrave Macmillan 2018), with Michele Zappavigna; and a special issue of *Functions of Language* (2018) focusing on interpersonal grammar. Eight volumes of his collected papers (edited by Wang Zhenhua) have been published in China (2010, 2012; Shanghai Jiao Tong University Press). In April 2014, Shanghai Jiao Tong University opened its Martin Centre for Appliable Linguistics, appointing Professor Martin as Director.

Karl Maton is Professor of Sociology at the University of Sydney and Director of the LCT Centre for Knowledge-Building, as well as Visiting Professor at the University of the Witwatersrand (South Africa) and Visiting Professor at Rhodes University (South Africa). Karl's book *Knowledge and Knowers: Towards a realist sociology of education* is the founding text of Legitimation Code Theory (Routledge, 2014). A collection of studies illustrating how to enact Legitimation Code Theory in research, *Knowledge-building: Educational studies in Legitimation Code Theory* (K. Maton, S. Hood & S. Shay, eds), was published by Routledge in 2016. Karl co-edited *Accessing Academic Discourse* (2020, Routledge) with J. R. Martin and Y. J. Doran, illustrating the productive dialogue between Legitimation Code Theory and systemic functional linguistics.

David Rose is Director of *Reading to Learn*, a literacy program that trains teachers across school and university sectors in Australia, Africa, Asia and Europe (www. readingtolearn.com.au). He is an Honorary Associate in the Dept of Linguistics at the University of Sydney. His research includes analysis and design of classroom discourse, pedagogic design for embedding literacy in curriculum learning and for beginning literacy, professional training for teachers in pedagogy and language, language typology, language evolution and social semiotic theory. Publications include *The Western Desert Code: An Australian cryptogrammar* (2001, Pacific Linguistics), *Genre Relations: Mapping culture* (with J. R. Martin, 2008, Equinox) and *Learning to Write, Reading to Learn: Genre, knowledge and pedagogy in the Sydney School* (with J. R. Martin, 2012, Equinox).

Nicky Wolmarans is a Senior Lecturer in the Department of Civil Engineering at the University of Cape Town. She holds a PhD from the University of Cape Town and an MSc in Mechanical Engineering. She teaches on the undergraduate program as well as having responsibility for academic development. She also teaches on the postgraduate programs specializing in Engineering Education Research. Her research interests focus mainly on understanding the nature of professional knowledge from a realist perspective.

1

THE TEACHING OF SCIENCE

New insights into knowledge, language and pedagogy

Y. J. Doran, Karl Maton and J. R. Martin

Introduction

Science is significant. Faced with the climate emergency, global pandemics and proliferating threats to life on Earth, that significance should go without saying. However, science is under sustained attack by irrationalism in politics, the news and social media. The need for science education has never been more urgent. Yet, in many advanced societies, science is struggling to attract and retain students through school and university. One problem facing attempts at addressing these issues is a tendency for studies to obscure the knowledge and language that comprise science. The subfields of education research dedicated to science disciplines have contributed greatly to our understanding of *scientific ways of knowing*. Studies explore the cognitive resources, perceptions and judgements of students and the processes of learning. They are extensively examining the conceptions, motivations and dispositions that students bring to learning science and the ways of thinking exhibited by students when learning science. However, *what* students are learning when they study science, the nature of the *scientific knowledge* itself, receives far less attention. Moreover, this dominant focus on how students learn science has been accompanied by neglect of the teaching of science. Pedagogy is too often reduced to an afterthought of findings about how students think. Without understanding how scientific knowledge and language may help shape the best ways of teaching that knowledge, we have only part of the picture. This volume aims to help fill this gap by exploring the knowledge practices and multimodal discourses of science teaching.

The collection extends two approaches that make pedagogic discourse a central object of study in education: Legitimation Code Theory (LCT) and Systemic Functional Linguistics (SFL). These approaches bring to light the meaning-making activities of social actors, in complementary ways. LCT is a sociological framework that foregrounds the knowledge practices of science education, revealing features such as their complexity, context-dependence, boundedness and specialized procedures. SFL explores the language and other semiotic resources, including mathematics, images and gesture, through which these knowledge practices are expressed. In recent years scholars and educators using these two frameworks have been closely working together, generating a fast-growing body of work that draws on both sides of this genuinely interdisciplinary dialogue (e.g. Martin and Maton 2013, Martin *et al.* 2020b). This book builds on this collaboration to offer cutting-edge developments in both approaches that will generate major advances in not only how science education is understood but also how knowledge, language and pedagogy are conceived more widely.

Teaching Science is organized into three parts. Part I draws on LCT to explore how science teaching can support knowledge-building. A series of innovative studies focus on the integration of mathematics into science (Chapter 2), the building of scientific explanations (Chapter 3) and the integration of multimedia such as animations into science teaching (Chapter 4). These studies introduce new concepts and new methods in LCT, including 'autonomy tours' (Chapter 2), 'constellation analysis' (Chapter 3) and 'epistemic affordances' (Chapter 4).

Part II greatly extends SFL to explore the multimodal discourses that underpin knowledge-building in science classrooms. These chapters articulate the expansive range of meanings involved in explaining complex scientific phenomena (Chapter 5), the ways in which deep scientific taxonomies are built (Chapter 6), and the essential interdependence of language, mathematics and images in scientific knowledge (Chapter 7). They advance the modelling of meaning-making in SFL into new areas and articulate extensions to concepts that will kickstart new forms of research into science discourse, including 'field' (Chapter 5), 'ideational discourse semantics' (Chapter 6) and the multimodal analysis of language, mathematics and image (Chapter 7).

Part III explores the practical implications that LCT and SFL analyses can deliver for teaching science. Chapters discuss how access to scientific knowledge can be widened to a greater diversity of students (Chapter 8), how knowledge is transformed to address real-world problems in engineering design (Chapter 9), the meanings communicated in live lectures that are rarely taken into account in debates over pedagogy (Chapter 10), and how mathematics can be taught effectively to all students through the influential pedagogic program 'Reading to Learn' (Chapter 11).

In this opening chapter we introduce the book by outlining the traditions of studies enacting the frameworks of LCT and SFL to examine science education. First, we locate this volume in the long-standing body of work using SFL to understand talking, writing and reading science. Second, we turn to the more recent but growing body of research and practice enacting LCT to examine and shape teaching and learning science. Finally, we introduce the chapters of this book, highlighting how they offer new ways of understanding, analyzing and shaping science teaching.

The language of science

Systemic Functional Linguistics (SFL) is a major linguistic approach that has developed worldwide since the 1960s. SFL explores how language is used in social life to organize our interpersonal relations, manage the content meanings we want to express and bring this together to make coherent text. Scholars using the approach consistently emphasize the need to study language in a way that contributes to improving the world. This emphasis on 'appliable' linguistics means SFL engages with a very wide range of contexts and uses of language. One of its longest engagements has been with scientific discourse. At first this was not to examine science itself but rather to understand English grammar (Huddleston *et al.* 1968). Since the 1980s, however, SFL engaged more deeply with scientific texts educationally, from the perspective of teaching literacy in science. At the time, researchers were dissatisfied with prevailing approaches to literacy that emphasized writing without explicitly modelling the language patterns that students need to learn. In response, SFL scholars began developing ways of explicitly teaching differences in writing across subject areas, an approach which became known as 'genre pedagogy' (Rothery 1989, Rose and Martin 2012).

However, to teach the specific language patterns used in different subjects, educators needed to know those patterns. This necessitated a large-scale descriptive effort to map language features across the disciplinary spectrum, much of which was funded by the Metropolitan East Disadvantaged Schools Program in Sydney, Australia. A key focus of this work was to analyze the language of science using concepts from across the SFL framework. An early concern was with text types and text structures of science, conceptualized in terms of *genre*. This work is illustrated by the book *Factual Writing* in which Martin (1985) presented descriptions of different genres in primary school scientific texts, including *reports* that generalize experiences about things, *procedures* that focus on how to make things happen, and *explanations* that explain things.[1]

Through the late 1980s and 1990s these models of genre in science were greatly expanded as SFL researchers engaged with later primary school, secondary school and workplaces associated with science and technology (e.g. Rose *et al.* 1992, Unsworth 1997a, 1997b, 1997c, Veel 1992, 1997). This work formed part of major innovations in SFL as a framework that transformed linguistic modelling of not only scientific discourse but also language more broadly. Many of these innovations are brought together in a series of landmark books on talking, writing and reading science. *Talking Science* (Lemke 1990) engaged in depth with language and meaning-making in science education for the first time. *Writing Science* (Halliday and Martin 1993) offered significant advances in the grammatical modelling of scientific language and insights into the nature of scientific meaning and technicality from the perspective of the register variable *field*. *Reading Science* (Martin and Veel 1998) collected papers significantly expanding the reach of SFL research into science, including popular science, environmental discourse, and both pedagogic and workplace settings, as well as exploring features of scientific discourse such as grammatical metaphor and multimodality. These books laid foundations for subsequent SFL research into science and fed back into the development of pedagogic programs through SFL. Many of these programs have aimed at designing classroom practices and resources for improving the teaching of both the content of any particular subject area as well as the literacy practices so necessary for organizing this content. This has included developing frameworks for teaching students language about language ('metalanguage') so they can more easily talk and reflect upon their reading and writing, and designing classroom practices that enable a gradual handover of scientific ways of reading and writing from teachers to students (Unsworth 2001, Christie and Derewianka 2008, Rose and Martin 2012, Derewianka and Jones 2016, Dreyfus *et al.* 2016, Humphrey 2017). These programs have been internationally influential, underpinning among other things, the current Australian literacy curriculum. Underpinning this book, the interaction between SFL research into scientific discourse and its pedagogic development highlighted three key areas of ongoing exploration: the grammar of scientific language, technical meanings in science and the multimodal nature of science.

The grammar of scientific language

In *Writing Science*, Halliday (1993a) highlighted significant differences in the language used in science compared to that of other registers. Scientific language, he summarized, involves:

- *interlocking definitions,* where technical terms are mutually defining;
- *technical taxonomies*, where terms are ordered into complicated layers of classification and composition;
- *special expressions*, where large multi-word constructions are technical, rather than simply individual words;
- *lexical density*, where there is a relatively high number of lexical words per clause;
- *grammatical metaphor,* where a grammatical structure is 'mismatched' with its meaning, such as when the clause 'how quickly cracks in glass grow' is reconstrued as the nominal group 'glass crack growth rate';
- *syntactic ambiguity*, where grammatical forms produced by grammatical metaphor lead to ambiguity in their meaning; and
- *semantic discontinuity*, where steps in the logic of reasoning or specific links connecting technical meanings are presumed or not made explicit.

These individual language features, Halliday points out, are regular occurrences in scientific texts and perform distinct functions in the organization of technical meaning. Through the history of scientific discourse, these grammatical features evolved in conjunction with scientific knowledge. To construe new meanings of scientific thought, scholars such as Chaucer, Newton, Maxwell and Darwin required new forms of English. Over time, these 'new forms of English' became the dominant motif for scientific texts (Halliday 1993b, 1993c, Banks 2008, 2017). However, when a number of these language features are used together in the same text, they can make scientific language difficult for learners – they are 'grammatical problems'.

In terms of linguistic modelling, the grammar of science raised questions regarding the nature of language in general. One key issue in particular was the nature of grammatical metaphor (Simon-Vandenbergen *et al.* 2003). Grammatical metaphors involve meanings being reconstrued in an 'incongruent' manner. For instance, in the example given by Halliday noted above, 'how quickly cracks in glass grow' is reconstrued as a grammatical metaphor through the nominal group 'glass crack growth rate'. Metaphors such as these were shown to be key in moving into the language of school science and other subjects (Derewianka 2003). But how one can show that something is a grammatical metaphor rather than a 'normal' realization of a meaning was less clear (Halliday 1998). For example, in biology the word 'phagocytosis' is a noun, yet it seems to construe an 'event' in some sense because it names the process whereby pathogens in the body are engulfed and destroyed. However, unlike 'glass crack growth rate', it cannot easily be turned back into a more 'congruent' form – it is rare, if indeed it happens at all, for biology textbooks to talk about 'phagocytosing' as a verb. These more theoretical and descriptive questions led to a refocus on how SFL could model semantics in relation to grammar. Among other things they influenced the highly elaborated model of ideational semantics put forward by Halliday and Matthiessen (1999), which has in turn laid a platform for the discourse semantic modelling of Hao (2020b, Chapter 6 of this volume) and has shed further light on how grammatical metaphors work (Hao 2020a, Martin 2021).

For educational research, this work gave a sense of the language patterns students needed to write. However, it was clear that many subject areas valued getting the technical 'content' meanings correct over other linguistic patterns. Although concerns for genre and grammar were important, this was only to the extent that they were used in service of the technical meanings of science. This raised the question of what a 'successful' text in science education is, and more broadly how to model scientific meanings in general.

Technical meanings in science

To understand how technical meanings are organized in science, the grammatical and genre perspectives on scientific language in SFL were coupled with a perspective of *field* within SFL. Field is a variable of register (following the stratal model of Martin, 1992) which is realized by ideational meaning in language and notionally organizes the 'content' meanings of discourse. In the 1980s, one of the key issues for SFL scholars was the large taxonomies of technicality in science with which students must engage (a field perspective on one of Halliday's 'grammatical problems'). This required renovating existing SFL descriptions of field to make a series of key distinctions. First, taxonomies were modelled as being either of classification, relating things in terms of type–subtype, or of composition, relating things in terms of part–whole relations. Different text types emphasized different types of taxonomy, and as Wignell *et al.* (1993) argued in *Writing Science*, different fields could also be distinguished by exploring their taxonomies. For example, in the specialized field of bird watching, taxonomies of birds tend to be more elaborated than those typically used in everyday life. However, these taxonomies share the fact they both tend to be based on observable characteristics of birds, such as colour, markings, size, habitat, etc. In contrast, the scientific taxonomies of ornithology, in addition to being significantly more elaborated, are based on shared lineage shown through genetic comparison and requiring specialist equipment, rather than observable characteristics. This renovated model enabled texts to be explored in terms of the field-specific meanings they presented. However, a key issue that remained was how language is used to *build* these taxonomies. Chapter 6 of this volume (Hao) takes a major step forward by laying out an extended model of 'ideational discourse semantics' for understanding how highly elaborated taxonomies are built in biology.

A second distinction concerned the intricate sequences of events known as *activities* that often occur in explanations and procedures. In contrast to many fields, activities in scientific fields are often linked by 'implication' where an activity necessarily implicates another activity. An example of this from Wignell *et al.* (1993) concerns *convective uplift*:

> Convective uplift
> Air in contact with a warm surface will become heated and expand causing it to rise. Dew point will be reached, condensation will take place and convectional clouds will form.

This text presents a series of activities that are related by implication, where one necessarily follows another (here ^ indicates sequence):

> Air in contact with a warm surface will become heated
> ^
> and expand
> ^
> causing it to rise
> ^
> Dew point will be reached
> ^
> condensation will take place
> ^
> and convectional clouds will form

Each step in this series necessarily occurs because of the previous step. This contrasts with series of activities in genre such as stories where events tend to occur in a less definite fashion – things can 'go wrong' and unexpected events may occur (indeed this is one of the key features of stories, Martin and Rose 2008). As the text above shows, these activities may include technical terms (dew point, condensation, convectional clouds) that may be positioned within the taxonomies of scientific discourse, while the whole sequence itself is named technically as 'convective uplift'. Such naming of activities, alongside elaborated taxonomies and extensive use of grammatical metaphor, was shown in *Writing Science* to underpin the enormous technicality of science.

This model of field has informed decades of research into scientific discourse and has proven both a relatively intuitive means of understanding technical meaning and one that can be linked with language patterns. However, it still did not account for many areas of scientific meaning, including the various gradable properties (size, weight, force, etc.) that often distinguish items in taxonomies and in later years of secondary school are often realized through mathematics and graphs, nor how the various aspects of field (taxonomy and activity) could be integrated into a coherent 'whole' to explain or describe phenomena. Taking these issues as its point of departure, Doran and Martin (Chapter 5, this volume) directly build upon this model of field to encapsulate a wider range of scientific meaning-making that has been highlighted by research in recent years.

Multimodality in science

Running parallel to studies of grammatical, genre and field-based attributes of science has been a focus on the *multimodal* nature of its discourse. Books such as *Reading Images* (Kress and van Leeuwen 1990) opened the way for linguists to engage with semiotic resources beyond written language. The work of Kress and van Leeuwen (1996) on scientific diagrams, for example, showed that images were organized in highly conventional ways that could be analyzed in terms comparable to both the field and grammatical modelling being developed for language. This highlighted the importance of multimodal meaning making in science education and in broader schooling contexts, and heavily influenced the turn towards 'multiliteracies' in education. School literacy is no longer considered just reading and writing language but also viewing, drawing and organizing images, and a range of other meaning-making systems (New London Group 1996, Unsworth 2001, Kress *et al.* 2001). Understanding how individual semiotic resources enable this to happen became a key question for SFL multimodality researchers. Work on images along these lines has continued steadily since the 1990s, accelerating recently in relation to the role graphs play in science (Doran 2018a, 2019, Chapter 7 of this volume), the significant degree of meaning organized into scientific diagrams (Unsworth 2020, Martin *et al.* 2021, Doran 2019), the use of certain types of image as technical formalisms across disciplines (Yu in press, Doran 2020), and how images are dynamized into animations for teaching science (He 2020).

In addition to images, the advent of multimodality encouraged further developments for other semiotic resources. For science a key step forward in this regard was O'Halloran's (2005) work on mathematics, which began to fill a major gap in our understanding of scientific discourse. Along with work by Lemke (1998, 2003), this work emphasized that multimodal texts were more than the 'sum of their parts'. Although each semiotic resource organized its own meaning, these meanings were 'multiplied' (Lemke 1998) when used with other semiotic resources in a text to make new meanings not available to any individual resource. O'Halloran's study put forward for the first time an operationalizable description of how different semiotic resources of language, mathematical symbolism and images worked not only individually but also in terms of their interactions or 'intersemiosis'. This focus on intersemiosis has laid the groundwork for ongoing work in systemic functional work on science education. For example, Doran's (2018a, 2018b) model of mathematical symbolism built on O'Halloran's work to develop a fully systematized grammatical description of mathematical symbolism used in secondary school and university physics. It also explored the interaction between language and mathematics in terms of the genres these two resources realized together. This in turn raised theoretical questions about the modelling of semiotic resources outside of language. In particular Doran questioned the general assumption that the three metafunctions – ideational, interpersonal and textual – occur for all semiotic resources (e.g. Kress and van Leeuwen 1996, O'Halloran 2005). Doran (2018b) argued that when descriptions did not assume it, there was no evidence for identifying an interpersonal metafunction in mathematical symbolism. This work on symbolisms is currently being extended by Yu (in press), focusing on chemistry discourse, which is helping to open space for engagement with the highly technical texts of upper secondary school science. A number of chapters in this volume directly extend SFL work on multimodality, including Chapter 7 (Doran) on the interaction of mathematics, images and language in physics, Chapter 10 (Hao and Hood) on interactions between body language and language in chemistry lectures, and Chapter 11 (Rose) on teaching mathematics.

Alongside the aforementioned advances in understanding grammar and technicality, this move towards multimodality helped transform the SFL understanding of science between the late 1980s and early 2000s. New models of language and semiosis were generated that opened the way for new objects of study and drove new pedagogic applications. More recently, a further transformation has been taking place in how SFL approaches science education, one sparked by an ongoing dialogue between SFL and Legitimation Code Theory.

The knowledge of science

Legitimation Code Theory (LCT) is a framework for analyzing and shaping practices.[2] LCT integrates and extends insights from a range of influences but most explicitly builds on the sociological frameworks of Pierre Bourdieu and Basil Bernstein (see Maton 2014, 2018). Like SFL, LCT is concerned with the meaning-making activities of social actors. As its name implies, LCT provides tools for exploring the bases of legitimacy or 'rules of the game' in social fields, in ways that reveal the organizing principles underlying practice. Metaphorically, LCT gets at the DNA of practice.

Reflecting its sociological foundations, LCT views society as comprising relatively autonomous social fields of practice (such as law, medicine, education, etc.) that have distinctive resources and forms of status. In each social field, actors cooperate and struggle, both for more of what is viewed as signs of success and over what defines success in that social field. Their practices thus embody messages as to what should be dominant measures of achievement. This is to highlight that there is more to what we say or do than what we say or do. For example, if a teacher has a class undertake a practical experiment, they are teaching not only whatever scientific content the experiment imparts but also that engaging in concrete, tangible activities is important and that students discovering results by themselves is important. LCT reveals these kinds of messages by analyzing the organizing principles of practice.

These organizing principles can be manifold. Any set of practices has a diverse range of characteristics, such as their complexity, context-dependence, emphasis on specialized knowledge or personal experience, boundedness from other practices, and so forth. The organizing principles underlying practices are conceptualized by LCT as different species of *legitimation code*. The conceptual framework is structured into a series of *dimensions* or sets of concepts that each explore a distinctive species of legitimation code. There are currently three active dimensions – Specialization, Semantics and Autonomy – centred on exploring specialization codes, semantic codes and autonomy codes, respectively.[3] Put simply, Specialization explores how knowledge and knowers are articulated within practices, Semantics explores questions of context and complexity, and Autonomy explores where contents and purposes of practices come from. (On Specialization, see Maton 2014, 2020; on Semantics, see Maton 2013, 2014 and Maton and Chen 2020; on Autonomy, see Maton and Howard 2018, 2020, and Chapters 2 and 4 of this volume. For how LCT concepts relate together, see Maton 2016). These different dimensions do *not* refer to different sets of empirical practices but rather offer ways of revealing different organizing principles underlying the same set of practices. How many and which dimensions are drawn on in empirical research depends on the problem-situation (specific questions concerning a particular object of study).

LCT and science education

Scholars and educators are enacting LCT to examine and shape practices across the disciplinary map and in all kinds of educational institutions, as well as beyond education (e.g. Maton et al., 2016a; Winberg *et al.* 2021). The framework is widely applicable and studies of topics far beyond one's own substantive areas of concern can offer insights. For example, in developing the concept of 'constellations' to analyze the teaching of scientific explanations, Maton and Doran (Chapter 3, this volume) found constellation studies of ballet lessons, History courses and ethnopoetics all highly valuable. Thus, insights into science education offered by LCT research are not confined to studies expressly focused on that topic. Nonetheless, there is a growing body of work dedicated to exploring the teaching and learning of science using different dimensions of LCT either separately or in combination.

The dimension of Specialization is proving particularly valuable for exploring how knowledge and knowers come together in science education. For example, concepts from Specialization have opened up new ways of thinking about supporting 'epistemological access' to science knowledge for students from diverse social backgrounds (Ellery 2018, Chapter 8 of this volume). They are also providing practical tools for supporting engagement with decolonization in science education in ways that respect different ways in which knowledges and knowers are valued (Adendorff and Blackie 2021). Focusing on the 'knowledge' side, an interlocking series of studies led by Karin Wolff (e.g. Pott and Wolff, 2020, Wolff 2021) draw on the *epistemic plane* (Maton 2014: 171–195) to explore how teaching and learning emphasize in different ways the specialized procedures of engineering and/or the phenomena for which they are used. Combined with studies of issues from work-integrated learning (Winberg 2012) to student design projects for 'real world' problems (Wolmarans, Chapter 9, this volume), this work is building a sophisticated picture of engineering education.

The dimension of Semantics directly resulted from two major studies of teaching practices in secondary school science (see Maton 2020). Specifically, this work introduced the notion of *semantic waves*, which describes recurrent shifts in the context-dependence and complexity of knowledge, and the method of *semantic profiling* those changes over time (Maton 2013, Macnaught *et al.* 2013). These ideas are being successfully enacted across science education, including to support cumulative learning in chemistry (Blackie 2014), student transition from school to university biology (Mouton and Archer 2019), project-based learning in biology (Mouton 2019) and in chemistry (Veale *et al.* 2017), real-world applications in chemical engineering (Dorfling *et al.* 2019), student design practices in engineering (Wolmarans 2016), problem-solving in physics (Conana *et al.* 2021), and student assessments in chemistry (Rootman-le Grange and Blackie 2018) and physics (Georgiou *et al.* 2014; Steenkamp et al. 2019). Extending these concepts into exploring multimodality is also a burgeoning focus, such as the role of language, mathematics and image in physics (Doran 2018a).

Studies of science education have also played a crucial role in developing new concepts in the dimension of Autonomy. Research into teaching in science classrooms helped generate the notion of *autonomy tours* (Maton and Howard 2018), which shows how to successfully integrate science knowledge with other content and purposes, such as everyday experiences, metaphors, analogies and knowledge from other subjects. These are providing new ways of understanding how to successfully integrate mathematics and multimedia objects in science teaching (Maton and Howard, Chapters 2 and 4 of this volume).

Dialogue with SFL

A genuinely inter-disciplinary dialogue between LCT and SFL has been underway since the turn of the century. This collaboration built on and intensified previous discussions between Basil Bernstein (whose theory is a foundational framework for LCT) and Michael Halliday and Ruqaiya Hasan. This dialogue went through a series of phases (see Maton and Doran 2017). In the late 1990s, Bernstein's ideas were inspiring SFL scholars to think about knowledge structures in education (Martin *et al.* 2020a). The emergence of LCT in the early 2000s offered to SFL, among other things, a means of engaging more empirically with knowledge practices, in both research and teaching. This began a series of more intense phases of direct inter-disciplinary collaboration between LCT and SFL scholars. One result is a rapidly growing number of papers, books and doctoral theses that use both frameworks together to generate greater explanatory power (e.g. Martin *et al.* 2020b; Maton *et al.* 2016). In this volume, for example, Doran (Chapter 7) engages with both the LCT dimension of Semantics and the SFL register variable *field* to examine how language, image and mathematics come together in school physics. Another result is an ongoing series of innovations in each framework. For example, the LCT concepts of 'semantic gravity' and 'semantic density', which explore the context-dependence and complexity of knowledge practices, were in part stimulated by dialogue with SFL scholars. In turn, these LCT concepts provoked the development of new SFL concepts of 'presence' (Martin and Matruglio 2020) and 'mass' (Martin 2020), which bring together the many ways these issues are manifested in language. These new concepts have begun to show great promise in gathering together manifold linguistic resources in studies of science education (Hood and Hao, Chapter 10, this volume).

More generally, this dialogue with LCT has evoked in SFL research a growing interest in the role of linguistic and semiotic resources in knowledge-building. This focus is resulting in major theoretical advances. One recent example is Hao's development of ideational discourse semantics (2018, 2020a, 2020b), as illustrated by Chapter 6 of this volume. Building on Halliday and Matthiessen's (1999) description of ideational semantics, Hao has developed a model for exploring the ways that technical knowledge is built through unfolding discourse. This addresses an issue that had long vexed SFL scholars, with previous attempts at understanding discourse semantics regularly becoming blurred into either grammatical description or field-based description. Hao's model (Chapter 6) offers a clear and distinct descriptive level that enables a view of scientific language on its own terms. A second example, which complements Hao's work, is a newly extended model of field by Doran and Martin, as shown in Chapter 5 of this volume. Taken together with the expansive grammar of English put forward by Halliday (Halliday and Matthiessen 2014) and the models of genre developed for science (see Martin 1985, Martin and Rose 2008), this new period of research has enabled SFL to greatly expand our understanding of the rich and multifaceted language and multimodality inherent in science.

On *Teaching Science*

This volume offers major steps forward in how both LCT and SFL conceive science and science education. The title consciously echoes past landmark works in SFL, with *Teaching Science* capturing the principal focus of and stimulus for this new work. The subtitle, *Knowledge, language, pedagogy,* sets out the organization of the book into three main parts that explore: knowledge-building in teaching science; the multimodal discourses that underpin this knowledge-building; and how to improve the teaching and learning practices of science.

Part I comprises three chapters that develop new ideas in LCT to understand major issues in teaching science: the integration of mathematics (Chapter 2), the teaching of scientific explanations (Chapter 3) and the use of multimedia digital resources (Chapter 4). These chapters build on one another. **Chapter 2** (Maton and Howard) addresses the vexing question of how mathematics can be successfully integrated in science teaching. That students often struggle with mathematics in science lessons, even when they have little difficulty with those ideas in mathematics lessons, has been a long-running concern for educators. One reason this issue remains unsolved is that existing approaches cannot systematically distinguish 'mathematics' knowledge from 'science' knowledge. Maton and Howard introduce cutting-edge tools from the LCT dimension of Autonomy that enable knowledge practices to be distinguished without lapsing into either essentialist definitions that neglect how these bodies of knowledge differ between contexts or relativist claims that they are nothing but endless flux. The concepts are illustrated through detailed analyses of real-world classroom practices that show a key attribute of successful integration of mathematics in science teaching to be *autonomy tours* that shift between different contents and purposes in particular ways. The ideas outlined here are poised to have a major impact on both research and practice in education, far beyond science teaching.

Chapter 3 (Maton and Doran) focuses on the role played by relations among ideas in teaching scientific explanations. Reflecting their knowledge-blindness, dominant approaches to researching science education neglect the ways in which ideas are connected to create explanations. Maton and Doran introduce the method of *constellation analysis* from LCT as a way of revealing these relations among ideas. This innovative method is used to analyze explanations of the tides and seasons. In each case the logic of explanations presented in school textbooks is analyzed and compared to how the explanation is taught in a classroom. These analyses show that explanations of seemingly similar kinds of phenomena differ in terms of how ideas are related together and that the logic of these relations impacts on how they are taught in classrooms. Constellation analysis offers a new analytic method with huge potential as a practical tool for researchers, curriculum designers, educators and students.

Chapter 4 (Maton and Howard) examines how multimedia such as animations can be integrated into teaching science. Existing research overwhelmingly focuses on developing principles for designing multimedia that support cognitive processing of information. This chapter meets an urgent need to foreground teaching and knowledge as a first step towards developing pedagogic principles for teaching multimedia that support learning science. To do so, the chapter extends existing limited uses of the notion of 'affordances' to examine how the knowledge practices expressed by multimedia relate to those central to specific classroom tasks. These *epistemic affordances* are revealed through an innovative form of autonomy analysis that shows how the diverse elements of complex multimedia objects relate to the contents and purposes of specific classroom tasks. In-depth analyses of two contrasting examples of science teaching with animations show the pedagogic work required of teachers to integrate such multimedia and how LCT offers a way of getting to grips with these complex objects in real-world contexts.

Part II comprises detailed studies of the language and multimodal discourse of science teaching using newly-developed tools from Systemic Functional Linguistics. These chapters articulate the expansive range of meanings involved in explaining complex scientific phenomena (Chapter 5), the ways deep scientific taxonomies are built (Chapter 6), and the interdependence of language, mathematics and images in building scientific knowledge (Chapter 7). These three chapters push the modelling of language and meaning-making in SFL into new territory and articulate theoretical principles that will enable new forms of research into science discourse. They complement the LCT analyses of Part I to significantly expand our understanding of scientific knowledge and semiosis and how these are taught.

Chapter 5 (Doran and Martin) introduces an evolving model of the register variable 'field' for understanding the intricate explanations built in science. This explores how science can view phenomena from a static perspective in terms of elaborated taxonomies or a dynamic perspective in terms of unfolding activities. Doran and Martin also explore how large swathes of gradable properties can be arrayed and measured, and how all of these can be reconstrued and interconnected for any particular 'topic' in science. This description takes a big step forward in building upon decades of modelling of field in SFL and works to be generalizable across language, image, mathematics and a wide range of other semiotic resources.

This new model of field is being developed in dialogue with a new model of ideational discourse semantics that informs **Chapter 6** (Hao). This explores scientific meaning as it unfolds in text to create an intricate discourse semantic framework for grasping the elaborate taxonomies built in biology, and the various entities they marshal. Chapter 6 forms part of a larger discourse semantic model developed by Hao (2018, 2020a, 2020b) that significantly pushes forward knowledge of how scientific language organizes its technical meaning. Together with Chapter 5, this offers for the first time an expanded resource for modelling ideational meaning that links lexicogrammar, discourse semantics, field and genre.

Chapter 7 (Doran) explores scientific meaning as inherently multimodal. Focusing in particular on the interaction of mathematics, images and language in physics, Doran shows how each builds technical meaning in ways that can 'hand over' this meaning to others. Drawing on the SFL model of field developed in Chapter 5 and the dimension of Semantics from LCT, this chapter shows how semiotic resources are brought together to move from relatively common-sense meanings to technical, while at the same time moving between the empirical and the theoretical.

Part III exemplifies the close connection between research and practice that characterizes LCT and SFL research into science teaching. These chapters draw on LCT and/or SFL to explore how more students can be supported to access scientific knowledge (Chapter 8), the vexed problem of simplifying 'real world' problems to teach professional reasoning in engineering education (Chapter 9), the role of body language in face-to-face chemistry lectures (Chapter 10), and methods for teaching mathematics to enable all students to learn (Chapter 11).

Chapter 8 (Ellery) examines a science foundation course that is intended to support more students to successfully engage with learning science at university. Enacting the Specialization dimension of LCT to explore curriculum, staff beliefs and student experiences, Ellery shows that the course involves two different bases of achievement. One basis requires students to demonstrate their understanding of scientific knowledge, the other requires students to be a particular kind of scientific knower. Moreover, students must be the right kind of knower in order to access the right kinds of knowledge. Ellery uses LCT to dig beneath the surface of curricular intentions and show how educational practices may contradict those intentions. The chapter poses a significant challenge to current thinking about how more students can be supported to access science in education.

Chapter 9 (Wolmarans) problematizes the widely-held view of professional education, such as engineering, as learning how to apply disciplinary science knowledge to 'real world' problems. Enacting concepts from the Semantics and Specialization dimensions of LCT, Wolmarans analyzes design projects in a civil engineering course that are intended to authentically mimic professional problems. Such projects require simplification of the problem for students. The chapter pushes LCT analysis to distinguish between reduced complexity of the knowledge required by students to solve a problem and reduced complexity of the problem itself. Crucially, the analysis shows that simplifying 'real world' projects requires careful negotiation between how the problem and the specialized scientific knowledge are simplified or else problems ensue for teaching professional reasoning. In short, the chapter shows how teaching engineering is more than simply 'applying' science knowledge to 'real world' problems.

Chapter 10 (Hood and Hao) explores the rich meanings made by body language in face-to-face lectures, drawing on a developing model of paralanguage in SFL (Cléirigh 2011, Hood 2011, Martin and Zappavigna 2019, Ngo *et al.* 2021). Focusing on university lectures in chemistry, Hood and Hao show the extensive means through which technical meaning is distributed across language and body language. In particular, they draw on Hao's model of ideational discourse semantics (Chapter 6, this volume) to detail how the scientific construction of the world in spoken language is regularly coupled with a more 'common-sense' construction in body language, thereby grounding the technical meanings of science. Coming at a time where classes are being increasingly moved online, this chapter offers key insights into what can be lost when we move away from face-to-face teaching.

Chapter 11 (Rose) extends the internationally influential SFL pedagogic programme 'Reading to Learn' into the teaching of mathematics. Rose begins from the straightforward yet significant observation that, when teaching, the symbols of mathematics are typically written down on the board but the procedure of *how to do* the mathematics is spoken aloud. The procedure is thus potentially lost to student at the moment of being taught. To address this, Rose lays out a principled pedagogic programme that makes explicit the procedures involved in solving mathematical problems, exemplified through a secondary school lesson on trigonometry. This involves a series of tasks that progressively build written procedures for doing mathematics and gradually hands over control from teacher to students. As with all aspects of the Reading to Learn programme, this method is developed to ensure that all students have access to mathematical knowledge in ways that are accessible and deliverable in the classroom.

The chapters of this collection will launch new research agendas in understanding knowledge, language and pedagogy in science. They offer major leaps forward in both LCT and SFL, with implications far beyond teaching science.

Notes

1 This book also laid out models of a range of argumentative expository genres, recounts and descriptions, which are more typically found in the humanities and social sciences.
2 For LCT papers, blogs and events, see: www.legitimationcodetheory.com.
3 A fourth dimension, Temporality, is under fundamental redevelopment.

References

Adendorff, H. and Blackie, M. (2021) 'Decolonizing the science curriculum: When good intentions are not enough', in C. Winberg, S. McKenna and K. Wilmot (eds) *Building Knowledge in Higher Education*, London: Routledge, 237–54.

Banks, D. (2008) *The Development of Scientific Writing: Linguistic features and historical context*, London: Equinox.

Banks, D. (2017) *The Birth of the Academic Article: Le Journal des Sçavans and the Philosophical Transactions, 1665–1700*, London: Equinox.

Blackie, M. (2014) 'Creating semantic waves: Using Legitimation Code Theory as a tool to aid the teaching of chemistry', *Chemistry Education Research and Practice*, 15: 462–9.

Christie, F. and Derewianka, B. (2008) *School Discourse*, London: Continuum.

Cléirigh, C. (2011) '*Gestural and postural semiosis: A systemic functional linguistics approach to "body language"*', Unpublished manuscript.

Conana, H., Marshall, D. and Case, J. (2021) 'A semantics analysis of first year physics teaching: Developing students' use of representations in problem-solving', in C. Winberg, S. McKenna and K. Wilmot (eds) *Building Knowledge in Higher Education*, London: Routledge, 162–79.

Derewianka, B. (2003) 'Grammatical metaphor and the transition to adolescence', in A. Simon-Vandenbergen, M. Taverniers and L. J. Ravelli (eds) *Grammatical Metaphor*, Amsterdam: John Benjamins, 185–220.

Derewianka, B. and Jones, P. (2016) *Teaching Language in Context*, Melbourne: Oxford University Press.

Doran, Y. J. (2018a) *The Discourse of Physics: Building Knowledge through Language, Mathematics and Image*, London: Routledge.

Doran, Y. J. (2018b) 'Intrinsic functionality of mathematics, metafunctions in systemic functional semiotics', *Semiotica*, 225: 457–87.

Doran, Y. J. (2019) 'Building knowledge through images in physics', *Visual Communication*, 18(2): 251–77.

Doran, Y. J. (2020) 'Academic formalisms: Toward a semiotic typology' in J. R. Martin, Y. J. Doran and G. Figueredo (eds) *Systemic Functional Language Description*, London: Routledge, 331–58.

Dorfling, C., Wolff, K. & Akdogan, G. (2019) 'Expanding the semantic range to enable meaningful real-world application in chemical engineering', *South African Journal of Higher Education*, 33 (1): 42–58.

Dreyfus, S., Humphrey, S., Mahboob, A. and Martin, J. R. (2016) *Genre Pedagogy in Higher Education: The SLATE project*, Basingstoke: Palgrave Macmillan.

Ellery, K. (2018) 'Legitimation of knowers for access in science', *Journal of Education*, 71: 24–48.

Georgiou, H., Maton, K. and Sharma, M. (2014) 'Recovering knowledge for science education research: Exploring the "Icarus effect" in student work', *Canadian Journal of Science, Mathematics, and Technology Education* 14(3): 252–68.

Halliday, M. A. K. (1993a) 'Some grammatical problems in scientific English, in M. A. K. Halliday and J. R. Martin (eds) *Writing Science*, London: Falmer, 76–94.

Halliday, M. A. K. (1993b) 'On the language of physical science' in M. A. K. Halliday and J. R. Martin (eds) *Writing Science*, London: Falmer, 59–75.

Halliday, M. A. K. (1993c) 'The construction of knowledge and value in the grammar of scientific discourse', in M. A. K. Halliday and J. R. Martin (eds) *Writing Science*, London: Falmer, 95–116.

Halliday, M. A. K. (1998) 'Things and relations: Regrammaticizing experience as technical knowledge', in J. R. Martin and R. Veel (eds) *Reading Science*, London: Routledge, 185–236.

Halliday, M. A. K. and Martin, J. R. (1993) *Writing Science: Literacy and Discursive Power*, London: Falmer.

Halliday, M. A. K. and C. M. I. M. Matthiessen (1999) *Construing Experience Through Meaning: A Language-Based Approach to Cognition*, London: Cassell.

Halliday, M. A. K. and C. M. I. M. Matthiessen (2014) *Halliday's Introduction to Functional Grammar*, London: Routledge.

Hao, J. (2018) 'Construing scientific causality in published research articles in biology', *Text and Talk*, 38(5): 520–50.

Hao, J. (2020a) 'Nominalisation in scientific English: A tristratal perspective', *Functions of Language*, 27(2): 143–73.

Hao, J. (2020b) *Analysing Scientific Discourse from a Systemic Functional Perspective*, London: Routledge.

He, Y. (2020) '*Animation as a Semiotic Mode: Construing Knowledge in Science Animated Videos*', unpublished PhD Thesis, University of Sydney.

Hood, S. (2011) 'Body language in face-to-face teaching: A focus on textual and interpersonal metafunctions', in S. Dreyfus, S. Hood and M. Stenglin (eds) *Semiotic Margins*, London: Continuum, 31–52.

Huddleston, R. D., Hudson, R. A., Winter, E. O. and Henrici, A. (1968) *Sentence and Clause in Scientific English*, London: UCL.

Humphrey, S. (2017) *Academic Literacies in the Middle Years*, London: Routledge.

Kress, G. and van Leeuwen, T. (1990) *Reading Images*, Geelong: Deakin University Press.

Kress, G. and van Leeuwen, T. (1996) *Reading Images*, 2nd edition, London: Routledge.

Kress, G., Jewitt, C., Ogborn, J., and Tsatsarelis, C. (2001) *Multimodal Teaching and Learning: Rhetorics of the Science Classroom*, London: Continuum.

Lemke, J. L. (1990) *Talking Science: Language, Learning, and Values*, Norwoord, N.J.: Ablex.

Lemke, J. L. (1998) 'Multiplying meaning: Visual and verbal semiotics in scientific text', in J. R. Martin and R. Veel (eds) *Reading Science*, London: Routledge, 87–113.

Lemke, J. L. (2003) 'Mathematics in the middle: Measure, picture, gesture, sign and word', in M. Anderson, A. Sanez-Ludlow, S. Zellweger and V. Cifarelli (eds) *Educational Perspectives on Mathematics as Semiosis*, Ottawa: Legas, 215–34.

Macnaught, L., Maton, K., Martin, J.R. and Matruglio, E. (2013) 'Jointly constructing semantic waves: Implications for teacher training', *Linguistics and Education*, 24(1): 50–63.

Martin, J. R. (1985) *Factual Writing: Exploring and Challenging Social Reality*, Geelong: Deakin University Press.

Martin, J. R. (1992) *English Text: System and Structure*, Amsterdam: John Benjamins.

Martin, J. R. (2020) 'Revisiting field: Specialized knowledge in secondary school science and humanities discourse', in J. R. Martin, K. Maton and Y. J. Doran (eds) *Accessing Academic Discourse,* London: Routledge, 114–47.

Martin, J. R. (2021) 'Ideational semiosis: A tri-stratal perspective on grammatical metaphor', *DELTA Documentação de Estudos em Linguistica Teorica e Aplicada* 36:3.

Martin, J. R. and Maton, K. (2013) (eds) 'Cumulative knowledge-building in secondary schooling', *Special Issue of Linguistics and Education,* 24(1): 1–74

Martin, J. R. and Matruglio, E. (2020) 'Revisiting mode: Context in/dependency in Ancient History classroom discourse', in J. R. Martin, K. Maton and Y. J. Doran (eds) *Accessing Academic Discourse,* London: Routledge, 89–113.

Martin, J. R. and Rose, D. (2008) *Genre Relations: Mapping culture,* London: Equinox.

Martin, J. R. and Veel, R. (eds) (1998) *Reading Science: Critical and Functional Perspectives on Discourses of Science,* London: Routledge.

Martin, J. R. and Zappavigna, M. (2019) 'Embodied meaning: A systemic functional perspective on paralanguage', *Functional Linguistics,* 6(1), 1–33.

Martin, J. R., Maton, K. and Doran, Y. J. (2020a) 'Academic discourse: An inter-disciplinary dialogue', in J. R. Martin, K. Maton and Y. J. Doran (eds) *Accessing Academic Discourse,* London, Routledge, 1–31.

Martin, J. R., Maton, K. and Doran, Y. J. (eds) (2020b) *Accessing Academic Discourse: Systemic functional linguistics and Legitimation Code Theory,* London: Routledge.

Martin, J. R., Unsworth, L. and Rose, D. (2021) 'Condensing meaning: Imagic aggregations in secondary school science', in G. Parodi (Ed.) *Multimodality,* Berlin: Peter Lang.

Maton, K. (2013) 'Making semantic waves: A key to cumulative knowledge-building', *Linguistics and Education,* 24(1): 8–22.

Maton, K. (2014) *Knowledge and Knowers: Towards a realist sociology of education,* London: Routledge.

Maton, K. (2016) 'Starting points: Resources and architectural glossary', in K. Maton, S. Hood and S. Shay (eds) *Knowledge-building,* London: Routledge, 233–43.

Maton K. (2018) 'Thinking like Bourdieu: Completing the mental revolution with Legitimation Code Theory', in J. Albright, D. Hartman and J. Widin (eds) *Bourdieu's Field Theory and the Social Sciences,* London: Palgrave Macmillan, 249–68.

Maton, K. (2020) 'Semantic waves: Context, complexity and academic discourse', in J. R. Martin, K. Maton and Y. J. Doran (eds) *Accessing Academic Discourse,* London, Routledge, 59–85.

Maton, K. and Chen, R. T.-H. (2020) 'Specialization codes: Knowledge, knowers and student success', in J. R. Martin, K. Maton and Y. J. Doran (eds) *Accessing Academic Discourse,* London, Routledge, 35–58.

Maton, K. and Doran, Y. J. (2017) 'SFL and code theory', in T. Bartlett and G. O'Grady (eds) *Routledge Systemic Functional Linguistic Handbook,* London: Routledge, 605–18.

Maton, K. and Howard, S. K. (2018) '*Taking autonomy tours: A key to integrative knowledge-building',* LCT Centre Occasional Paper 1: 1–35.

Maton, K. and Howard, S. K. (2020) Autonomy tours: Building knowledge from diverse sources, *Educational Linguistic Studies,* 2: 50–79.

Maton, K., Hood, S. and Shay, S. (eds) (2016a) *Knowledge-building: Educational Studies in Legitimation Code Theory,* London, Routledge.

Maton, K., Martin, J. R. and Matruglio, E. (2016b) 'LCT and systemic functional linguistics: Enacting complementary theories for explanatory power', in K. Maton, S. Hood and S. Shay (eds) *Knowledge-building,* London, Routledge, 93–113.

Mouton, M. (2019) 'A case for project based learning to enact semantic waves: Towards cumulative knowledge building', *Journal of Biological Education,* DOI: 10.1080/00219266. 2019.1585379.

Mouton, M. & Archer, E. (2019) 'Legitimation Code Theory to facilitate transition from high school to first-year biology', *Journal of Biological Education*, 53(1): 2–20.

New London Group (1996) 'A pedagogy for multiliteracies', *Harvard Educational Review*, 66(1): 60–93.

Ngo, T., Hood, S., Martin, J. R., Painter, C., Smith, B. and Zappavigna, M. (2021) *Modelling Paralanguage Using Systemic Functional Semiotics*, London: Bloomsbury.

O'Halloran, K. (2005) *Mathematical Discourse: Language, symbolism and visual images*, London: Continuum.

Pott, R. and Wolff, K. (2020) 'Using Legitimation Code Theory to conceptualize learning opportunities in fluid mechanics', *Fluids*, 4(203): 1–13.

Rootman-le Grange, I. & Blackie, M. (2018) 'Assessing assessment: In pursuit of meaningful learning', *Chemistry Education Research and Practice*, 19: 484–90.

Rose, D. and Martin, J. R. (2012) *Learning to Read, Reading to Learn: Genre, knowledge and pedagogy in the Sydney School*, London: Equinox.

Rose, D., McInnes, D. and Körner, H. (1992) *Scientific Literacy*, Sydney: Metropolitan East Disadvantaged Schools Program.

Rothery, J. (1989) 'Learning about language', in R. Hasan and J. R. Martin (eds) *Language Development: Learning language, learning Culture*, Norwood, NJ: Ablex, 199–256.

Simon-Vandenbergen, A., Taverniers, M. and Ravelli, L. (eds) (2003) *Grammatical Metaphor: Views from Systemic Functional Linguistics*, Amsterdam: John Benjamins.

Steenkamp, C., Rootman-le Grange, I. and Müller-Nedebock, K. (2019) 'Analysing assessments in introductory physics using semantic gravity', *Teaching in Higher Education*, DOI: 10.1080/13562517.2019.1692335

Unsworth, L. (1997a) 'Scaffolding reading of science explanations: Accessing the grammatical and visual forms of specialized knowledge', *Reading*, 31(3): 30–42.

Unsworth, L. (1997b) 'Explaining explanations: Enhancing scientific learning and literacy development', *Australian Science Teachers Journal*, 43(1): 34–49.

Unsworth, L. (1997c) '"Sound" explanations in school science: A functional linguistics perspective on effective apprenticing texts', *Linguistics and Education*, 9(2): 199–226.

Unsworth, L. (2001) *Teaching Multiliteracies Across the Curriculum*, Buckingham: Open University Press.

Unsworth, L. (2020) 'Intermodal relations, mass and presence in school science explanation genres' In M. Zappavigna and S. Dreyfus (eds) *Discourses of Hope and Reconciliation: J. R. Martin's Contribution to Systemic Functional Linguistics*. London: Bloomsbury, 131–152.

Veale, C., Krause, R. and Sewry, J. (2017) 'Blending problem-based learning and peer-led team learning, in an open ended "home-grown" pharmaceutical chemistry case study', *Chemistry Education Research and Practice*, 19(1): 68–79.

Veel, R. (1992) 'Engaging with scientific language: A functional approach to the language of school science', *Australian Science Teachers Journal*, 38(4): 31–35.

Veel, R. (1997) 'Learning how to mean – scientifically speaking: Apprenticeship into scientific discourse in secondary school', in F. Christie and J. R. Martin (eds) *Genre and Institutions*, London: Cassell, 161–95.

Wignell, P., Martin, J. R. and Eggins, S. (1993) 'The discourse of geography: Ordering and explaining the experiential world', in M. A. K. Halliday and J. R. Martin (eds) *Writing Science*, London: Falmer, 151–83.

Winberg, C. (2012) '"We're engaged": Mechanical engineering and the community', in P. Trowler, M. Saunders, and V. Bamber (eds), *Tribes and Territories in the 21st Century*, London: Routledge, 142–55.

Winberg, C., McKenna, S. and Wilmot, K. (eds) (2021) *Building Knowledge in Higher Education: Enhancing teaching and learning with Legitimation Code Theory*, London: Routledge.

Wolff, K. (2021) 'From principle to practice: Enabling theory-practice bridging in engineering education', in C. Winberg, S. McKenna and K. Wilmot (eds) *Building Knowledge in Higher Education*, London: Routledge, 180–97.

Wolmarans, N. (2016) 'Inferential reasoning in design: Relations between material product and specialised disciplinary knowledge', *Design Studies*, 45(A): 92–115.

Yu, Z. (in press) 'Chemical formalisms in secondary school chemistry: Toward a semiotic typology', *Semiotica*.

PART I

Knowledge-building in science education

2

TARGETING SCIENCE

Successfully integrating mathematics into science teaching

Karl Maton and Sarah K. Howard

... like trying to hit a bullet with a smaller bullet, whilst wearing a blindfold, riding a horse.
— 'Scotty', in *Star Trek*, 2009

Introduction

In his *Opus Majus* of 1267, Roger Bacon described mathematics as 'the door and key' to science. This has become an axiom of educational research into science and its constitutive disciplines.[1] Mathematics is widely heralded as the 'backbone' of science (Bing and Redish 2009: 1) and 'deeply woven' into its practice and teaching (Redish and Kuo 2015: 562). Science textbooks are shown to exhibit greater use of mathematics than those of other disciplines (Lemke 1998, Parodi 2012). Learning the 'appropriate application' of mathematical skills is said to be 'a key part of the hidden curriculum in science' (Quinnell *et al.* 2013: 814). Accordingly, the ability to integrate mathematical and scientific knowledge is viewed as an important sign of student progress (Redish 2017). In short, teaching and learning mathematical knowledge is a central issue for science education. However, just as widely acknowledged is that integrating mathematics into science teaching poses persistent problems. Studies regularly proclaim that many students struggle with mathematics in science lessons and consequently become discouraged from continuing in science (Meli *et al.* 2016). Even students who have chosen further studies in science 'often seem to see "maths" as a separate subject, a necessary evil ... rather than an integral part of the discipline' (Quinnell *et al.* 2013: 811).

A key aspect of the problem is said to lie with fundamental differences between the two disciplines. Scholars emphasize that mathematics when used within science has 'a different purpose – representing meaning about physical systems rather than expressing abstract relationships. It even has a distinct semiotics ... from pure mathematics.' (Redish and Kuo 2015: 563). Integrating mathematics into science education is thus not a simple matter of adding mathematical content to science lessons. Indeed, differences between the disciplines are such that students encountering a concept located within both bodies of knowledge often fail to recognize that they are exploring similar ideas in different contexts (e.g. Planinic *et al.* 2012). As a result, students may succeed in mathematics classes but 'fail to use those same tools effectively' in science classes, leaving educators 'distressed and confused' (Redish 2017: 25).

A challenge facing science education is thus to identify and develop teaching practices that select, recontextualize and integrate mathematical knowledge in ways that support the learning of scientific knowledge. Unsurprisingly, this has been the subject of a significant body of research. The resulting work has generated suggestive ideas for learning specific mathematical skills for particular scientific problems. However, there is no overarching or integrating model for successful pedagogic integration. The problem 'remains unsolved' (Redish and Kuo 2015: 561). This chapter contributes towards the creation of such a model. It does so through by offering a fresh approach that complements existing frameworks by bringing to light issues that have hitherto been sidelined.

We begin by highlighting how research into science education offers significant insights into how students learn scientific ways of knowing but either neglects the forms of knowledge being taught or, where knowledge is discussed, treats 'science' and 'mathematics' as self-evident and unchanging. We argue that these constructions of the problem help underpin its persistence by ignoring knowledge, backgrounding teaching, and glossing over how 'science', 'mathematics' and their interrelations vary across contexts and change through the course of education. We then outline a framework that can complement existing approaches by bringing these issues into the picture. We introduce concepts from the Autonomy dimension of Legitimation Code Theory that conceptualize one aspect of the forms taken by knowledge practices and allow research to capture changing relations between changing forms of knowledge. Specifically, *autonomy codes* reveal the organizing principles underlying different knowledge practices, *autonomy pathways* trace changes in relations between knowledge practices over time, and *targets* embrace the contextual nature of what is viewed as 'science' and 'mathematics'. We illustrate the value of these concepts in analyses of science classrooms drawn from a major study of secondary schooling. These show that different autonomy pathways taken by teachers enable or constrain the integration of mathematics into classroom science. We conclude by reflecting on the potential of the concepts of *autonomy codes*, *pathways* and *targets* to bring knowledge into the picture and to connect specific instances of pedagogic practice together within a general model of pedagogic integration.

Studying mathematics in science: blind spots

To integrate mathematics into classroom science requires teaching practices that appropriately select ideas from one body of knowledge (mathematics) and recontextualize that selection within a second selection of ideas from another body of knowledge (science). A key issue is thus how different teaching practices shape the forms taken by ideas from those bodies of knowledge when they are brought together. Put simply, the question is: what teaching practices enable or constrain the integration of mathematical knowledge into scientific knowledge? These statements may seem unnecessary. However, studies of science education typically sideline both teaching practices and changing forms of knowledge. These blind spots arise from three assumptions that pervade the field: that knowledge equates to knowing, that education equates to learning, and that 'science' and 'mathematics' are self-evident.

Knowing and learning

The first assumption is that 'knowledge' comprises mental processes of under-standing that reside 'in the heads of persons' (von Glasersfeld 1995: 1). Reflecting this 'subjectivist doxa' (Maton 2014: 3–14), research focuses almost exclusively on cognitive and affective ways of knowing. Knowledge as an object of study in its own right – one taking particular forms which have effects for bringing that knowledge together with other forms – is left out of the picture. Put another way, the assumption is that to analyze knowledge one must analyze ways of knowing. Rather than distinguish between students' dispositions and what they are learning, as a precursor to exploring relations between knowing and knowledge, the only concern is the former. This assumption that knowledge is nothing but knowing is typically accompanied by a second assumption: that education is nothing but learning. When studying 'ways of knowing', research overwhelming focuses on student interactions, such as when solving a scientific problem. Teaching is rarely centre stage, if considered at all. Each of these assumptions thereby takes part of the picture for the whole.

This focus on learning and ways of knowing is illustrated by studies using the 'resources framework' (e.g. di Sessa 1993, Hammer 2000, Redish 2014, 2017), an influential approach to physics education research. The framework explores 'how our students think' (Redish 2014), such as 'the student's perception or judgement (unconscious or conscious) as to what class of tools and skills is appropriate to bring to bear in a particular context' (Bing and Redish 2009: 1). The concern is how 'cognitive resources' are 'activated in response to a perception and interpretation of both external and internal contexts' (Redish and Kuo 2015: 573). Thus student perceptions are central – what their perceptions may be about, the forms taken by knowledge itself, is not analyzed. This subjectivism is thoroughgoing – everything is psychological. For example, the term 'epistemology' is used to refer not to inter-subjectively shared field-level knowledge practices but rather to personal frames of individual understanding (e.g. di Sessa 1993, Hammer and Elby 2002). Accordingly, disciplines such as mathematics are viewed as comprising 'ways of knowing' (Redish 2017) and studies of mathematics in science explore how students solve problems in order 'to model their thinking' (Bing and Redish 2009: 2).

Research using the 'resources framework' offers valuable insights into how students learn ways of knowing. However, learning is not the sum of education, and ways of knowing are not the sum of disciplines. Much remains missing from the equation. Such studies could provide a powerful basis for understanding the integration of mathematics into science lessons if they were complemented by analyses of teaching and analyses of knowledge. However, frameworks for bringing these issues into the picture are lacking in the wider field. Currently, whatever approach they are using, studies of science education tend to reduce knowledge to knowing and education to learning, generating blind spots in the overall field of vision.[2] For example, studies of mathematics in science education draw on such frameworks as 'thinking dispositions' to suggest that attributes such as curiosity can help students shift from 'rigidity of mind' to 'fluid thinking' that ostensibly supports successful integration (Quinnell *et al.* 2013). Similarly, studies adopting a 'cognitive blending framework' analyze how students draw on 'mental spaces' when combining physical and mathematical knowledge (Bing and Redish 2007). Thus, everything lies in the mind of the beholder. Similarly, teaching is sidelined. The implications of studies of student learning for how mathematics should be taught in science often take the form of afterthoughts, as if teaching is merely an epiphenomenon of learning. Typically, such implications simply comprise calls for the integration of mathematics into science in teaching (e.g. Meli *et al.* 2016; Planinic *et al.* 2012), leaving unsaid what teaching practices would support that integration. In short, the widely shared focus on ways of knowing (rather than also knowledge) and on learning (rather than also teaching) limits current understanding of how pedagogic practices enable or constrain the integration of mathematics into science within classrooms.

Knowledge as self-evident and invariant

Knowledge is not entirely absent from discussions of mathematics in science. As mentioned earlier, differences between their purposes and 'semiotics' are said to contribute to student difficulties. Typically, science is described as condensing additional meanings into numbers and symbols and as requiring a different approach to interpreting mathematical results, reflecting its relationship with the external world (e.g. Bing and Redish 2009, Redish and Kuo 2015). Such attributes are also highlighted in discussion of the 'affordances' of mathematical representation for science (e.g. Fredlund *et al.* 2012). In systemic functional linguistics, work has shown how mathematics 'multiplies' meanings in relation with language and images (Lemke 1998), offering additional resources for construing scientific knowledge (O'Halloran 2010). More recently, Doran (2018) has brought together systemic functional linguistics with Legitimation Code Theory (LCT) to identify key mathematical genres in school physics and explore the role these play alongside language and images as physics progresses through school. Studies using LCT on its own have also explored forms of scientific knowledge (including mathematical symbols) involved in teaching and assessment in terms of differences in their complexity and context-dependence (Georgiou *et al.* 2014; Blackie 2014; Conana *et al.* 2016).

What is left open, however, is the question of identifying 'mathematics' and 'science'. To analyze integration in a classroom requires determining which knowledge is 'mathematics', which knowledge is 'science', and when a specific idea or practice has been recontextualized from one into the other. However, existing studies typically describe content, such as the formula or problem being discussed by students, and simply state what is 'mathematical' and what is 'scientific' (e.g. Meli *et al.* 2016), as if self-evident. Alternatively, discussions of the nature of 'science' or 'mathematics' make claims about each discipline as a whole, as if homogeneous and unchanging. That these ways of constructing knowledge are problematic flows from two uncontentious commonplaces. First, however distinctive the bodies of knowledge populating their *intellectual fields* might be (and this is debatable), the manifestations of 'science' or 'mathematics' *in a specific classroom* cannot be assumed. Which knowledge from an intellectual field is selected, recontextualized and enacted as curricula, and which knowledge from a curriculum is selected, recontextualized and enacted in classroom pedagogy varies geographically, institutionally and through the stages of education.[3] Put simply, the knowledge practices comprising 'science' and 'mathematics' are not necessarily the same in two classes in a school, let alone in different schools, years of study, states or countries. Second, at what stage of education specific knowledge practices from 'mathematics' are integrated into and become 'science' is not universal. What has already been integrated into 'science' in one classroom may remain separate 'mathematics' in another classroom. In short, what is 'science' varies between classroom contexts and changes through education, the 'mathematics' drawn on when teaching 'science' varies and changes, and the degree to which that 'mathematics' has been transformed into 'science' also varies and changes. Classroom 'science' and 'mathematics' are two variable and changing bodies of knowledge whose interrelations are themselves situational and mutable. They are anything but self-evident, homogeneous or unchanging – thus our liberal use of quote marks. To simply state that specific ideas are 'science' or 'mathematics' is to beg the question of how that is determined.

Addressing the question is not easy. It requires avoiding a false dichotomy between essentialism and relativism that continues to bewitch education research (Maton 2014: 1–22). On the one hand, universalizing claims about 'science' and 'mathematics' without a limiting context (such as 'science in this classroom') can lead to essentialism that treats their properties as homogeneous and invariant. On the other hand, insistence on the contextual limits of any definitions (or offering no more than 'science in this classroom, at this moment') can slide into relativism that treats 'science' or 'mathematics' as an endless flux. The former generates overgeneralized claims that are unhelpful for analyzing empirical data; the latter leads to the banal conclusion that subject areas are constructed, contested and fluid, paralyzing the possibility of analysis. As yet, research into science education has not steered a course between this Scylla and Charybdis. However, without facing squarely the question of distinguishing knowledge practices, empirical studies can only continue creating a series of context-bound models of specific instances. The wider issue of pedagogic integration will remain unsolved.

Seeing into the blind spots

Assuming that knowledge is only knowing, that education is only learning, and that disciplinary knowledges are self-evident generates blind spots. It is difficult to develop pedagogic practices that integrate mathematics into science lessons so long as teaching practice and both forms of knowledge are not analyzed. Existing insights into how students learn ways of knowing thus need to be complemented by: (i) studies of teaching practice; (ii) concepts that make visible the forms of knowledge practice being taught and learned; and (iii) a means of enacting those concepts that captures the variant and contextual nature of 'science' and 'mathematics'.[4] The first requires a shift of empirical focus. The second can be addressed by drawing on Legitimation Code Theory, a framework that reveals the organizing principles of knowledge practices. The third is trickier – it is akin to how Scotty described transwarp beaming in the motion picture *Star Trek*: 'like trying to hit a bullet with a smaller bullet, whilst wearing a blindfold, riding a horse'. It needs to account for two changing phenomena (what is 'science' and what is 'mathematics') whose relations are also changing (through different degrees of separation and integration). We now turn to a means for doing so.

Autonomy

Legitimation Code Theory or 'LCT' is a framework for researching and shaping practice. It begins from the notion that there is more to what we say or do than what we say or do. In other words, the meanings of practices are not exhausted by their content; practices are also 'languages of legitimation' or criteria for measuring achievement (Maton 2014). In short, what we say or do express principles of legitimacy or 'legitimation codes'. LCT comprises several *dimensions* or sets of concepts that explore different aspects of legitimacy (Maton 2016). Central to each dimension are concepts for analyzing the organizing principles underlying practices, dispositions and contexts as a particular species of 'legitimation code'. In terms of our needs here, these concepts bring knowledge into the picture by revealing the organizing principles generating its various forms. The dimension most directly relevant to exploring integration is Autonomy, which focuses on relations between sets of practices (such as subject areas) and conceptualizes their organizing principles as *autonomy codes*. We shall first define the concepts, then discuss how they are enacted using *translation devices* and *targets*. For reasons that become clear, we begin rather abstractly, before concretizing the meanings of concepts.

Autonomy codes

The dimension of Autonomy begins from the simple premise that any set of practices comprises constituents that are related together in particular ways. Constituents may be actors, ideas, institutions, machine elements, body movements, etc.; how they are related together may be based on explicit procedures, tacit ways of working, mechanisms, unstated orthodoxies, etc. The concepts of 'autonomy codes' explore how practices distinguish their constituents and their ways of relating from those of other practices. Put another way, the concepts examine how practices establish different degrees of insulation around their constituents and the ways those constituents are related together. These are analytically distinguished as:

- *positional autonomy* (PA) between constituents positioned within a context or category and those positioned in other contexts or categories; and
- *relational autonomy* (RA) between the relations among constituents of a context or category and the relations among constituents of other contexts or categories.

Each may be stronger (+) or weaker (−) along a continuum of strengths, where stronger represents greater insulation and weaker represents lesser insulation. Stronger positional autonomy (PA+) indicates that constituents of a context or category are relatively strongly delimited from constituents associated with other contexts or categories (strongly insulated positions); and weaker positional autonomy (PA−) indicates where such distinctions are less demarcated (weakly insulated positions). Stronger relational autonomy (RA+) indicates that the ways of relating constituents together are relatively specific to a set of practices (autonomous principles), and weaker relational autonomy (RA−) indicates that the ways of relating may be drawn from or shared with other sets of practices (heteronomous principles).

As shown in Figure 2.1, positional autonomy and relational autonomy are visualized as axes of the *autonomy plane*. Varying their strengths independently (PA+/−, RA+/−) generates four principal autonomy codes:

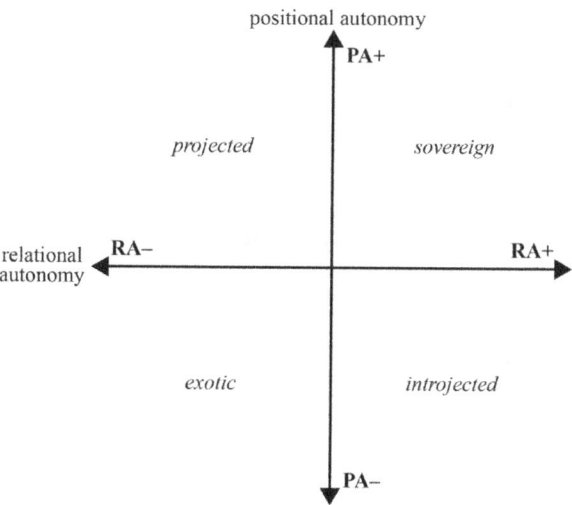

FIGURE 2.1   The autonomy plane (Maton and Howard 2018: 6)

- *sovereign codes* (PA+, RA+) of strongly insulated positions and autonomous principles, where constituents are associated with the context or category and act according to its specific ways of working;
- *exotic codes* (PA−, RA−) of weakly insulated positions and heteronomous principles, where constituents are associated with other contexts or categories and act according to ways of working from other contexts or categories;
- *introjected codes* (PA−, RA+) of weakly insulated positions and autonomous principles, where constituents associated with other contexts or categories are oriented towards ways of working emanating from within the specific context or category; and
- *projected codes* (PA+, RA−) of strongly insulated positions and heteronomous principles, where constituents associated with the specific context or category are oriented towards ways of working from elsewhere.

These concepts help address the need to make visible the forms of knowledge being taught and learned. Put simply, the four codes state that what matters are: internal practices and principles (sovereign codes); other practices and principles (exotic codes); other practices turned to intrinsic purposes (introjected codes); and internal practices turned to other purposes (projected codes). To explore processes that occur through time, such as classroom practice, one can analyze the different *pathways* traced around the plane by successive autonomy codes. There is an unlimited number of potential pathways (see Maton and Howard 2018). In this chapter we discuss the two pathways illustrated in Figure 2.2: *one-way trips* that begin in one code and end in another code; and *tours* that begin in one code, move through other codes, and return to their originating code. We shall show that autonomy tours in teaching practice enable, and one-way trips constrain the integration of 'mathematics' into 'science'. However, before doing so there remains the question of defining

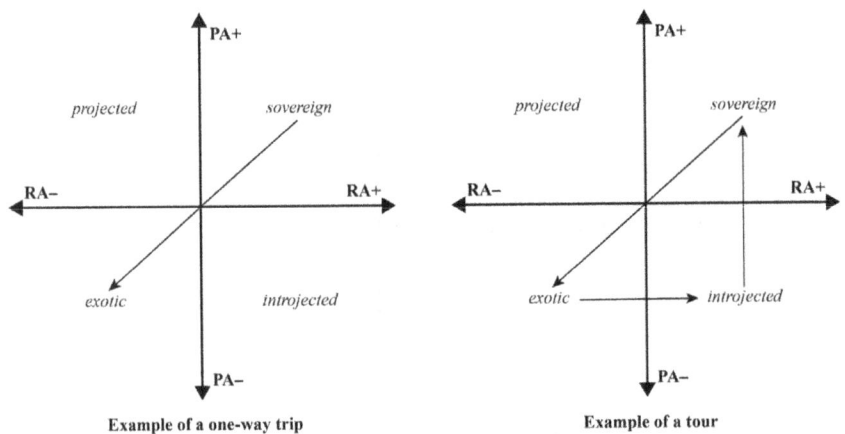

FIGURE 2.2 Examples of two autonomy pathways

'science' and 'mathematics' in a way that systematically embraces their variant and contextual nature. This is achieved through *translation devices* and *targets*.

Translation devices and targets

Thus far we have described 'autonomy codes' in abstract terms and without specific examples. This allows the concepts to be enacted across a wide range of diverse phenomena. Doing so offers the possibility of reaching beyond descriptions of specific instances of classroom practice to generate a general model of pedagogic integration. However, it also means one must be clear how the concepts are manifested within a specific object of study. In LCT this is achieved through 'translation devices' that relate concepts to data (Maton and Chen 2016). Table 2.1 is a *generic translation device* that relates autonomy codes to all forms of data. The device divides the continua of strengths for positional autonomy and relational autonomy into categories of progressively finer-grained levels of delicacy, from categories for stronger/weaker (*target/non-target*) through subcategories, use of which depends on the analysis.

To activate the device one asks: what constituents (practices, beliefs, ideas, actors, etc.) and what principles (purposes, aims, ways of working, etc.) are considered constitutive of *this* context or category, here, in *this* space and time, for *these* actors? This gives a 'target'. As shown in Table 2.1, target constituents embody stronger positional autonomy and all other, non-target constituents embody weaker positional autonomy; similarly, target principles embody stronger relational autonomy and all other, non-target principles embody weaker relational autonomy. These categories can be divided into subcategories by asking which target constituents and principles are considered *core* and which *ancillary* to the context or category, and which non-target constituents and principles are considered *associated* or *unassociated* with the target. Asking the same basic questions again generates a third level comprising *inner* and *outer* forms of core and ancillary targets, and *near* and *remote* forms of associated and unassociated non-targets.

TABLE 2.1 Generic translation device (Maton and Howard 2018: 10)

PA/RA	1st level	2nd level	3rd level
+	target	core	inner
			outer
		ancillary	inner
			outer
	non-target	associated	near
			remote
		unassociated	near
−			remote

Activating the device allows us to aim directly at the problem of defining 'science' and 'mathematics'. Rather than shying away from knowledge practices varying across contexts, the notion of 'target' makes that the starting point for analysis. This is best shown by a concrete example. Here we draw on a major study of how secondary school teachers select, assemble and enact knowledge in their classroom practice when teaching at Stage 4 (Years 7–8) and Stage 5 (Years 9–10) in three secondary schools in New South Wales, Australia.[5] Data comprised videorecordings of lessons across whole units of study (6–8 hours each), interviews with teachers, all teaching materials, and student artefacts. In this chapter we shall discuss examples of classroom practice by two science teachers in Year 7 of secondary schooling.[6]

To enact 'target' first one considers *whose* view of the target to begin from, as other agents in a context (such as students in a classroom) may have different targets. A target is always *someone's* or *something's* conception of what makes a context distinctive and thus in our analyses always accompanied by possessives (e.g. his/her/their targets). Here, reflecting our concern with teachers' practices, we focus on their targets. Second, one must consider *what level* of their targets to examine. LCT concepts can be enacted at all levels of analysis; for example, a teacher's targets may include an entire curriculum, a unit of study, a lesson, a task, and so forth. Reflecting our concern with how teachers attempt to integrate mathematics into science to meet the needs of the Stage 4 curriculum, our specific translation device has curriculum stage as its first level and unit of study as its second level.[7] As summarized in Table 2.2, in interviews and pedagogic materials the teachers identified their target content (PA+) as the Stage 4 science syllabus in New South Wales and their target purpose (RA+) as teaching students that content. Put simply, here: *positional autonomy* conceptualizes where the ideas expressed in classroom practice are drawn from, the Stage 4 science syllabus (PA+) or elsewhere (PA–); and *relational autonomy* conceptualizes the purposes for which they being expressed, teaching and learning that science syllabus (RA+) or other purposes (RA–). Interviews and pedagogic materials further identified the teachers' *core* targets (++) as the specific science unit being taught, with other units in Stage 4 science considered *ancillary* targets (+).

TABLE 2.2 Simplified specific translation device for this analysis

PA/RA	1st level	In this analysis:	2nd level	In this analysis:
+	target	New South Wales Stage 4 syllabus for subject area	core	specific unit in target
			ancillary	other topics or years in target
	non-target	other contents or purposes	associated	other educational knowledge
–			unassociated	knowledge from beyond education

(Their *inner-core* targets comprised the content points they created for each specific lesson). In terms of non-targets, teachers viewed other educational knowledge (such as other subjects or other Stages and levels of education in science) as *associated* (–) with their target, and knowledge from beyond education as more distanced or *unassociated* (– –).

By 'targeting' analysis we can identify 'science' as it is constructed in the specific context under study, avoiding universalizing essentialism. By translating that particular set of empirical ideas and practices into 'autonomy codes', we can move beyond context-bound, endlessly varying descriptions of difference, avoiding relativism. We can *both* embrace the specificities of each context *and* compare practices across different contexts, capturing the endless forms most wonderful that are 'science'. Moreover, targeting 'science' allows us to analyze the movement of ideas and practices between subject areas as they are recontextualized and integrated. Given that the 'target' depends on the object of study, no single idea, practice, belief, etc. is always and everywhere the same code. A practice may be moved around the plane; for example, in our discussion of an autonomy tour below, 'graphing' is successively constructed as an exotic code (as mathematics content for mathematical purposes), an introjected code (mathematics content for learning science), and a sovereign code (science content for learning science).

We can now begin to explore what teaching practices enable or constrain integration of 'mathematics' into 'science'. To illustrate how, we shall analyze classroom practices by two teachers (mentioned above) from Year 7 schools teaching the same unit from the same state curriculum. The difference between the examples lies in the autonomy pathways traced by their teaching practice. In the first example, the teacher fails to integrate mathematics into science. He leads students on a *one-way trip* out of 'science' into an activity he describes as 'maths' that remains segmented from his target knowledge. In the second example, a different teacher takes students on an *autonomy tour* that integrates non-target 'mathematical' knowledge about creating graphs into her target 'science' knowledge about Earth's seasons.

One-way trip from science 'to do some maths'

Our example of teaching that fails to integrate 'mathematics' into 'science' comprises a distinct phase of activity spanning an entire lesson of over 50 minutes. The teacher's core target for the wider unit, as later described in an interview, is:

> to teach them [students] about the universe and our solar system and what's beyond Earth. Some of them didn't quite understand the relationships in the universe so we have to make them clearer for them.... How we get night and day or how you get the different seasons.

This reflects a 'sub-strand' of the state curriculum for Year 7 science entitled 'Earth and space sciences', which is 'concerned with Earth's dynamic structure and its place in the cosmos' (NESA).[8] In the lesson discussed here, the teacher tells students they are going to make sense of the scale of the solar system, but quickly shifts classroom practice into using numbers to calculate percentages, detached from learning about the science content. This takes students on a *one-way trip* from the teacher's sovereign code into an exotic code, a pathway that does not return to his target content or target purpose. After 52 minutes he draws the 'maths' activity to a close by declaring: 'I know it's confusing'.

Pathway into 'a lot of numbers'

The teacher begins the lesson by showing students a short YouTube video entitled 'The smallest to the biggest thing in the universe'. Starting from hypothesized entities in quantum physics (such as strings), the video zooms outwards through ever-larger phenomena to end with the known galaxy. He then segues to the activity that will consume the rest of the lesson:

TEACHER   So, as you saw, some of those distances and some of those sizes don't really mean a lot to us, because we just can't fathom the distances involved, okay? So some of the other distances, especially in our solar system, are the same. So what we're going to do is, we're going to put the distances and the sizes relative to Earth. Okay? So we're going to put all the planets and the distance to the Sun and we're going to make them relative to the Earth.

At this point, the intended classroom practice is to explore content about the solar system (stronger positional autonomy) for the purpose of understanding the solar system (stronger relational autonomy); i.e. within the teacher's core target – deep inside his sovereign code.

The teacher then directs students to 'draw up a table' of 'seven columns and 10 rows' and shows a PowerPoint slide of a table, to which he adds two column titles by hand on the whiteboard, reproduced here as Table 2.3. He tells the class to 'copy down this information if you haven't already got it'. After reminding students they

TABLE 2.3 Table provided by teacher for activity

	Radius (km)	Distance from the sun (km)	Time to orbit around the sun	Time taken to turn once on its axis	Diameters as % of Earths	Distance as % of Earths
The Sun	695800					
Mercury	2439.7	57910000	88Ed	58d15h30m		
Venus	6052	108200000	224.7Ed	116d18h0m		
Earth	6371	149600000	365.25Ed	1d		
Mars	3390	227900000	686.97Ed	1d0h40m		
Jupiter	69911	778500000	12Ey	9h56m		
Saturn	58232	1433000000	29Ey	10h39m		
Uranus	25362	2877000000	84Ey	17h14		
Neptune	26422	4503000000	165Ey	16h06m		

had written down diameters of planets and their distances from the sun in a previous lesson, he explains:

TEACHER   I want you to add this information, these two [points to third and
fourth titled columns], because these two are relative to the Earth. … Since
we've already got the information just do four columns, because we're going
to do some maths.

That the teacher declares 'we're going to do some maths' does not by itself indicate
a shift beyond 'science' into non-target knowledge. For example, 'maths' could refer
to procedures or ideas he has previously integrated into his target – already scien-
tized mathematics, so to speak. Similarly, calculating percentages is not necessarily
beyond his target. As we shall see in our second example, no practice is always a
specific code. To identify the autonomy code, one must leave aside assumptions of
what is 'science' or 'maths' and begin from the teacher's target. This he described,
and teaching materials revealed, as the Stage 4 science syllabus. A strand of this
syllabus entitled 'science inquiry skills' includes for Year 7: 'Summarise data, from
students' own investigations and secondary sources, and use scientific understanding
to identify relationships and draw conclusions based on evidence'.[9] So, calculating
diameters and distances from the sun of planets as percentages of those of Earth
could *potentially* sit within the teacher's target. However, the syllabus emphasizes
that such 'science inquiry skills' give students 'the tools they need to achieve deeper
understanding of the science concepts' – they must be 'closely integrated' with
learning the 'science knowledge' outlined in a strand entitled 'Science understanding'. This strand includes having 'students view Earth as part of a solar system, which
is part of a galaxy, which is one of many in the universe, and explore the immense
scales associated with space'. Thus, whether calculating percentages lies within the
teacher's target depends on whether he integrates its content or purpose with viewing Earth as part of a solar system and exploring the immense scales of space. As
the teacher stated at the outset, this was his intention. However, as we shall see, in
practice he does not relate the activity to any such 'science understanding'.

Instead, as illustrated by Figure 2.3, the teacher quickly shifts the task into an
exotic code in which non-target content is used for non-target purposes. He weakens positional autonomy by disconnecting the contents of the table from his target
topic. For example, he describes its contents as 'information' five times in just the
first minute: 'copy down this information … just use your information … add
this information … you've already got the information … we've already got the
information'. At the same time he weakens relational autonomy by describing the
purpose as 'to do some maths' without relating this either to procedures previously
integrated into 'science' or to learning new syllabus content.[10] For example, when
responding to questions from students, he states:

TEACHER   Just do the last two columns and then add two more because we're
going to do some maths in the last two.

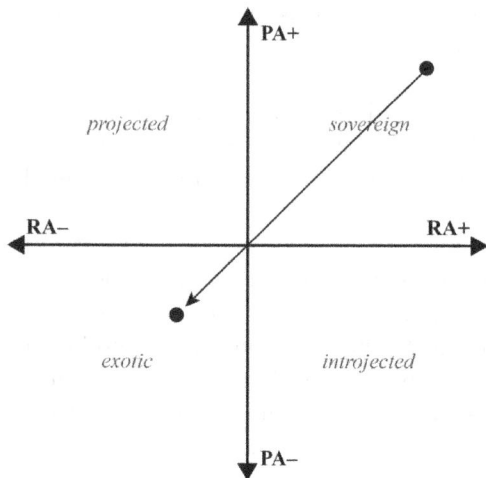

FIGURE 2.3 Shift from sovereign code to exotic code

Following his instructions, some students copy numbers from the table while a series of students ask the teacher which numbers they should copy. This concern with numbers as only numbers continues when, after eight minutes, he addresses the whole class:

TEACHER Alright, do we know how to work out the percentage for ....? So which one do we have to divide by? [Student name], what do you think? To work out Mercury, the percentage compared to Earth? What do you think we'd have to do?

STUDENT Divide it by a hundred?

TEACHER No, no, no. Alright. What we do [draws on whiteboard: $\frac{Mercury}{Earth} \times 100$]. Alright, so distance percentage [pointing to last column title] is the distance from the sun. Okay? So to work out the percentage, you divide each of the planets by the Earth's diameter.

Over the next 15 minutes the teacher repeats similar instructions to a series of individual students, each time describing what 'information' must be multiplied or divided to 'give you a percentage'. When he mentions the names of planets, the teacher is referring to specific empty cells in the table – shown by physically pointing to the cell – rather than to planets. More often, the teacher refers to the content as 'information' or 'number', such as:

TEACHER Divide that number [pointing to the table] by that number [pointing] for the distance; that number [pointing] by that number [pointing] for the diameter. Alright? That's what you're supposed to be doing.

The content is from neither the teacher's target of the syllabus nor his core target of learning about the solar system; the purpose is to 'work out the percentage' or 'do maths'. The knowledge being expressed thus remains within an exotic code (Figure 2.3).

During the course of the activity some students ask questions that represent opportunities to relate the activity to the issue of grasping the scale of the solar system. For example, one student asks whether people are made of 'planks', another asks what is smaller than a 'string' (both mentioned in the earlier video), and a third asks 'Are we made of stardust?'. The teacher's responses – 'No', 'Didn't you watch the video?', and 'What do you think?', respectively – do not connect to target content or turn the questions to his target purpose.

After 38 minutes the teacher asks students to call out numbers for cells in the column 'Diameters as % of Earths'. He then raises the question: 'What do these percentages actually mean?'. This is an opportunity to strengthen relational autonomy by turning these numbers to the purpose of viewing Earth as part of a solar system or exploring the immense scales involved, and an opportunity to strengthen positional autonomy by connecting the 'information' to knowledge about the solar system. A student suggests 'A lot of numbers', an answer that accurately reflects the exotic code characterizing the activity. The teacher leaves his own question unanswered. The class then repeats the pattern outlined above: students make calculations (for the 'Distances' column), the teacher repeats similar instructions to students, and numbers are solicited from the class. Classroom practice stays in an exotic code. The activity is ended after 52 minutes by the teacher saying 'I know it's confusing' and announcing that they will look at 'day and night' in the next lesson.

Summary: 'That's maths!'

The autonomy pathway traced by this lesson represents a one-way trip out of the teacher's target of 'science' in order 'to do some maths'. As portrayed by Figure 2.3, the knowledge expressed in classroom practice shifts from a fleeting sovereign code to a very long stay in an exotic code. As shown by the times given above, almost the entire 'science' lesson is 'maths'. The teacher could have chosen to conduct this activity inside his sovereign code by closely integrating the numeric activity with his syllabus target. Instead, he chooses to project the activity as beyond his target, as doing 'maths' to 'work out the percentage'. As we discuss below, this code shift is not necessarily antithetical to integrating this 'maths' into 'science'. At any point during the lesson, the teacher could strengthen positional autonomy by connecting to his target content or strengthen relational autonomy by turning non-target content (calculating percentages) to his target purpose. Instead, he keeps classroom practice in the exotic code: the content remains numbers and calculations, and the purpose remains using numbers to calculate other numbers. Thus, the shift to an exotic code does not integrate 'maths' into 'science'. Late in the lesson, in response to a student declaring 'This is hard, sir', the teacher replies 'That's maths! We still have to do maths in science'. However, this 'maths' is not 'in science' and so knowledge of calculating percentages remains strongly segmented from knowledge of the solar system.

Autonomy tour integrating 'mathematics' into 'science'

To illustrate how 'mathematics' can be integrated into 'science' we turn to a different teacher at a different school but teaching the same unit ('Earth and space sciences') at the same level (Year 7 secondary school). The example begins in the second lesson of a unit on the causes of Earth's seasons, as students transform their results from a practical experiment into graphs. In the first lesson students had conducted an experiment to explore the effect on temperature of the angle at which sunlight strikes the Earth's surface. In groups, students used a lamp to represent the sun, and a wooden block to represent the Earth. Varying the angle of the lamp to the block (15, 30, 60 and 90 degrees), they recorded the temperature of the block at different times (initial, 2.5 minutes, 5 minutes) from an attached thermometer. Prior to the experiment each student had written a hypothesis of whether increasing the angle would increase, decrease or have no effect on the temperature. The second lesson directly builds on this activity. The teacher begins by setting out her (inner-core) target:

TEACHER What we will be doing today is looking at those results, graphing the results and then talking about what it is that we were actually trying to model.

Over the next 35 minutes the teacher leads students on an autonomy tour through those activities: from her sovereign code (discussing their results), through an exotic code (recapping 'graphing rules'), and an introjected code (applying those rules to graph their results), before returning to her sovereign code (by relating the resulting graphs to Earth's seasons). As a result, the graphing activity becomes integrated into 'science'.

A tour through 'graphing'

The teacher begins by recounting the experiment and then solicits students' overall findings:

TEACHER So looking at your results there, who can give me a statement about what their results did?
STUDENT As the angle of the block increased, the temperature increased.
TEACHER Fantastic. I love that. That's a really great statement. Did someone get something different in their results?

The teacher thus begins deep inside her sovereign code. Both content (experiment modelling a factor in Earth's seasons) and purpose (to learn about the results) are located within her *inner-core target* for the lesson. After discussing the findings of several students, she announces:

TEACHER   Here [on the whiteboard] is your table that you should have had drawn up from the last lesson. We are going to graph... I want you to think about the graphing rules and start getting yourself ready for graphing.

As we emphasized, no activity is intrinsically a specific autonomy code. 'Graphing' is not necessarily non-target – 'graphing' can be mathematical or scientific. To determine autonomy codes we must consider the teacher's target: the Stage 4 science syllabus. A strand entitled 'science inquiry skills' includes for Year 7: 'Construct and use a range of representations, including graphs, keys and models to represent and analyze patterns or relationships in data'.[11] Thus, graphing is potentially within the teacher's target. However, as discussed in the previous example, the syllabus describes 'science inquiry skills' as giving students 'the tools they need to achieve deeper understanding of the science concepts' by being 'closely integrated' with learning the 'science knowledge' outlined in the syllabus strand 'Science understanding'. This strand includes 'how changes on Earth, such as day and night and the seasons, relate to Earth's rotation and its orbit around the sun'. Thus, whether graphing lies within the teacher's target depends on whether she integrates its content or purpose with learning about Earth's seasons. Here, we shall show that she begins by separating graphing in terms of both content and purpose, then turns it to purpose, before connecting its content.

 This tour begins with the teacher recapping her 'graphing rules' separately from Earth's seasons. Continuing on from the preceding classroom quote, she says:

TEACHER   So, who can remind me about what the rules are for graphing?
STUDENT   Y versus X.
TEACHER   Y versus X. How do we know which one goes where?
STUDENT   The independent variable goes on one side.
TEACHER   The independent variable goes on one of them. Yes, that's good.
STUDENT   And the dependent variable …
TEACHER   … goes on the other one. The thing that is the most regular, which is usually your IV [independent variable], goes on the X, and your DV [dependent variable] goes on the Y.

This recap embodies: weaker positional autonomy (PA–), as these 'graphing rules' are not related to Earth's seasons; and weaker relational autonomy (RA–), as the purpose is recapping the 'graphing rules' rather than learning about Earth's seasons. As portrayed in Figure 2.4, the teacher has shifted from deep inside her sovereign code to just inside an exotic code.[12]

 Thus far, classroom practice traces the same pathway as the previous example. However, where that teacher remained within an exotic code for the entire lesson, this teacher does not stay for long. She quickly shifts the class into a third code by repurposing the knowledge of 'graphing rules':

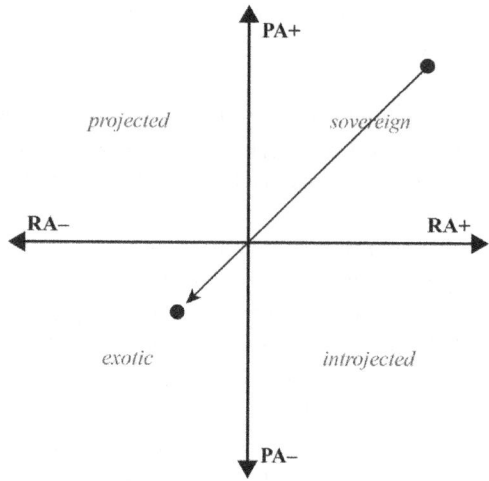

FIGURE 2.4 Shift from sovereign code to exotic code

TEACHER Now, in *this* experiment, who can tell me – there's a little problem. Have a look at our data. Can you tell me which one goes on the X and which one goes on the Y?

One student suggests 'the angle' should go on the Y-axis, another suggests the X-axis, and the teacher asks students for the locations of 'temperature' and 'time'. After a student exclaims 'Wait! What?', the teacher explains the problem:

TEACHER So in this experiment we've got three sets of data, okay? So, this one's going to kind of break the rules a tiny bit. The easiest way for us to do this is that you're going to have […] 'time' on the X, 'temperature' on the Y, and four different lines. The four lines you're going to draw is one line for 15 degrees, one line for 30 degrees, one line for 60 degrees and one line for 90.

The content of discussion – rules about locating variables on axes – remains weakly integrated with what the experiment reveals about Earth's seasons and so embodies weaker positional autonomy. However, the purpose is to create a graph that can show this knowledge, embodying stronger relational autonomy. As portrayed in Figure 2.5, this shifts classroom practice into an introjected code.[13]

This introjected code is maintained throughout the graphing activity. While students apply the adapted 'graphing rules' to their results, the teacher alternates between addressing the whole class and advising individual students; for example, to the class:

TEACHER All right! Along the X-axis, there will be three values: the X-axis has your time on it. There will be a time for five minutes, there will be a time for two and a half, and there will be a time for 'initial', which we can call zero, zero minutes. Okay? …

Then (continuing straight on), she looks at a student's workbook and asks:

TEACHER Why is this word here?
STUDENT 'Angle'.
TEACHER We are not doing 'angle' like that.
STUDENT Oh, whoops!
TEACHER Just follow what's going on here. This is the X.
STUDENT Okay.
TEACHER Okay? So 'temperature' does not belong there. X along here is 'time'. Y along here is 'temperature'.

Discussion continues along these lines for the next 12 minutes while students draw their graphs. As these quotes illustrate, the content of discussion involves locating variables on axes, setting ranges for variables, sizing the graph, using symbols, labelling, evenly spacing intervals, creating a key for symbols, and avoiding overlapping lines. Content is thus not related to what the results of the experiment might reveal about Earth's seasons: weaker positional autonomy. However, the purpose is to create graphs which help show what the experiment might reveal about Earth's season: stronger relational autonomy. As the teacher explains to the whole class: 'This is a better way of presenting the data than it is to look at a table. … Straight away when you look at this graph, you can see which one has increased in temperature fastest.' Graphing thus manifests here as an introjected code (Figure 2.5).

Once students have completed graphing, the teacher shifts classroom practice back to her sovereign code. Students write in their workbooks a 'conclusion' of

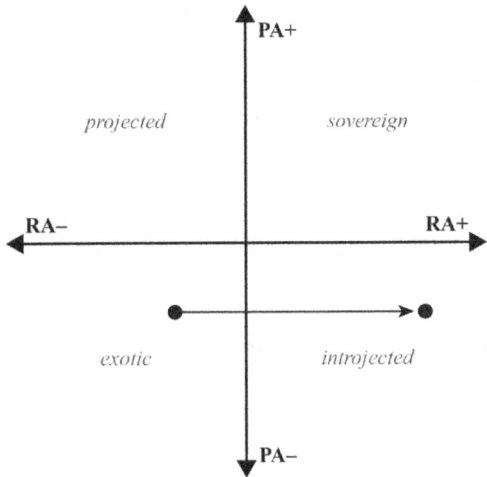

FIGURE 2.5 Shift from exotic code to introjected code

what their graphs show and whether this supports or refutes their previous hypothesis and a 'discussion' of whether their results were consistent and expected, what errors occurred, how they could improve their experimental design, and what they could do to test the idea further. The teacher then leads a discussion of the graphs that continues this concern with what they reveal about the focus of the experiment:

TEACHER   What did we learn? [...]
STUDENT   We learned that the steeper the angle, the hotter the temperature.
TEACHER   Good. The steeper the angle of the block, we got a greater increase in our temperature. Who can tell me why? Why did it get hotter? ...
STUDENT   Because the core of the block is closer to the light.
TEACHER   Good. The middle part of the block, as you increase the degrees, makes it closer to the light. Good.
STUDENT   Because it's getting more direct rays when it's on a higher angle as opposed to when it's on ...
TEACHER   Good. When we have a higher angle, we have more of those light rays striking the block, and those light rays then can heat up the block more effectively than the ones that are just skimming over the top.

This shifts the content and purpose to exploring the results of the experiment. The graphing activity has now been integrated with the experiment. The teacher then consolidates this sovereign code to integrate the experiment into discussion of Earth's seasons. First, she emphasizes her target purpose – stronger relational autonomy:

TEACHER   Okay, but what's the point in doing this? Are we really interested in whether or not blocks can heat up with a lamp?
STUDENT   No!
TEACHER   No? Who can remember the word I used to describe what this experiment was? Starts with an 'm'.
STUDENT   A model?
TEACHER   A model. Fabulous. This was a model. It was a model of the Earth and the sun.

Second, she explains differences between the model and reality and how those differences shape the experience of heat on Earth, content that embodies stronger positional autonomy:

TEACHER   Does the Earth change its angle?
STUDENTS   Yeah.
STUDENTS   No!
STUDENTS   It rotates.
TEACHER   It rotates – good. When the Earth rotates the angle changes. ... When the Earth rotates, we change the angle that the sunlight is striking the Earth.

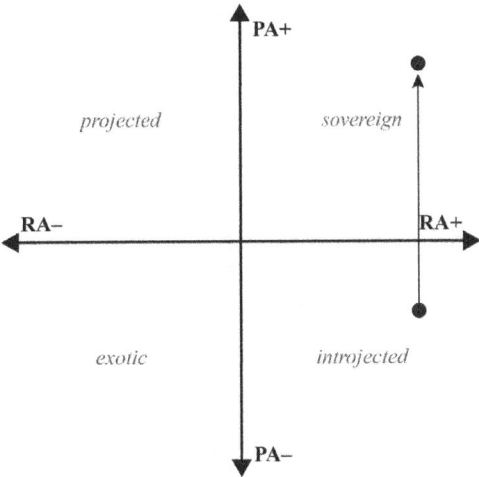

FIGURE 2.6 Shift from introjected code to sovereign code

The teacher then segues to an animation that shows the Earth rotating, sunlight striking the surface, and how this creates night and day. As portrayed in Figure 2.6, the teacher has shifted to content from the syllabus for the purpose of learning that syllabus – her sovereign code. The experiment and graphing activity are now integrated into the wider discussion of Earth's seasons.

Summary: separate, repurpose, integrate

Analysis of workbooks from this lesson suggests that students successfully translated their tables of experimental results into graphs and their graphs into conclusions about the effects of the angle of sunlight on temperature. This is no simple feat. Studies of science education widely report that many secondary school and university students struggle with understanding and interpreting graphs (Planinic *et al*. 2012). Indeed, the students here were *creating* graphs from data. Moreover, the teacher is also laying foundations for students' future learning. As she highlighted in an interview, her Year 7 students 'have no experience with graphing for science or they've got no experience with drawing tables for science – we've really got to teach that stuff in the beginning, because then we expect them to follow it through' subsequent years of school science.

In terms of the knowledge involved, this learning was supported by teaching which traced an autonomy tour from 'science' through graphing and back to 'science'. As shown by Figure 2.7, classroom practice went through:

(1) the teacher's *sovereign code* by discussing results of the experiment
(2) an *exotic code* when discussing 'graphing rules'

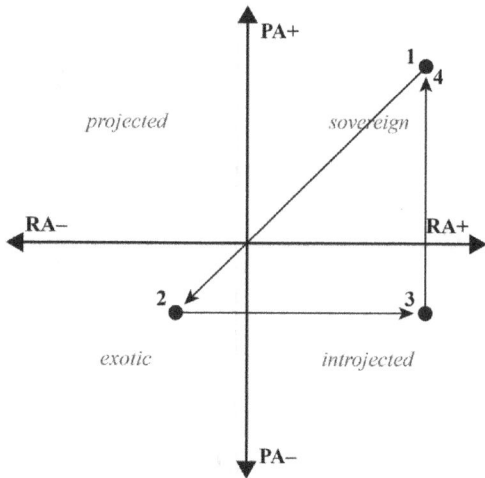

FIGURE 2.7 Autonomy tour with graphing results of an experiment

(3) an *introjected code* when adapting those 'rules' to graphing results of the experiment

(4) the teacher's *sovereign code* when translating graphs into conclusions about what was modelled by the experiment

Given that graphing could be discussed in ways that locate it either within or beyond the teacher's target, one question this pathway raises is why she chose to leave her sovereign code. Students had learned the 'graphing rules' in previous lessons, so the teacher could have treated them as part of her target – as already integrated into 'science' – by relating them directly to the experiment's result. However, this strategy would not have captured the distinct nature of this experiment – that there are three variables. By constructing the 'graphing rules' as an exotic code, the teacher keeps that knowledge separate from the specific 'science' content, allowing her to connect to ideas that students have already learned in a way that highlights how graphing will be different here. By separating this 'mathematics' from the 'science', she is able to select ideas from the 'graphing rules', repurpose those ideas, and integrate their use into her target knowledge. Separation comprises a shift to an exotic code. Repurposing involves strengthening relational autonomy by turning the 'rules' to the purpose of graphing results from this experiment – an introjected code. Integration involves strengthening positional autonomy to translate the resultant graphs into knowledge about Earth's seasons – a sovereign code. In our previous example, leaving 'science' for 'maths' was a one-way trip that failed to integrate the 'maths' back into 'science'. Here, leaving 'science' was a precursor to successful integration through an autonomy tour.

Conclusion

A key challenge faced by research into science education is developing teaching practices that select, recontextualize and integrate mathematical knowledge to support the learning of scientific knowledge. We argued that existing approaches offer insights into student learning of ways of knowing but typically sideline teaching practice and the forms of knowledge being taught and learned. To help address these blind spots we offered a complementary approach centred on *autonomy codes*, *pathways*, and *targets*. The concepts of *autonomy codes* foreground the forms taken by knowledge practices expressed in classroom discourse, focusing on a key relevant feature for integration: their relations with other knowledge practices. Analyzing the *pathways* traced by successive codes reveals how different teaching practices enable or constrain the integration of 'mathematics' into 'science'. Using these concepts, we analyzed contrasting examples of classroom practice from secondary school lessons in science that suggest *autonomy tours* support and *one-way trips* obstruct the integration of 'mathematics' into 'science'.

One implication of the analysis is that pathways to successful integration are not necessarily direct. In the tour example, the separation of 'mathematical' ideas from 'scientific' knowledge was an important precursor to its subsequent integration. This speaks to an issue highlighted by science education research: student difficulties with recognizing they are exploring mathematical ideas when presented in scientific contexts (e.g. Planinic *et al.* 2012). Constructing the 'mathematics' knowledge as separate (exotic code) is an opportunity to highlight the specific constellations of meanings within which that knowledge is located and which underpins its 'mathematical' nature. Turning those ideas to a 'scientific' purpose (introjected code) and connecting those repurposed ideas to 'scientific' knowledge (sovereign code) recontextualizes the ideas within a new constellation of meanings. A tour thus offers the possibility of making explicit those constellational differences. It makes the knowledge visible to students. As a growing body of research is showing, not all tours involve this specific combination of autonomy codes, but all involve departing and returning. If classroom discourse remained within a sovereign code throughout, these constellational differences would not be made visible; and if classroom discourse did not return, the recontextualization of ideas between constellations would not be possible.

While 'autonomy codes' help bring knowledge practices into the picture, we should emphasize that the concepts are not limited to that focus – one can also analyze students' dispositions, interactions and changing understandings. In short, the concepts can be enacted to examine *both* forms of knowledge practices and ways of knowing. This would enable 'matches' and 'clashes' to be identified, supporting the development of appropriate pedagogic practices.

Autonomy codes are, of course, not the only feature of knowledge practices, and autonomy tours are not the only factor in integration. As mentioned earlier, a key issue highlighted by physics education research is that science involves 'learning to blend physical meaning into mathematical representations and use that physical meaning in solving problems' (Redish 2017: 25). This can be traced through autonomy pathways: the second teacher condenses 'mathematical' ideas with empirical meanings when repurposing 'graphing rules' (from exotic code to introjected code) and when relating the resultant graphs to explaining Earth's seasons (from introjected code to sovereign code). In contrast, the 'maths' of the first teacher remains disconnected from empirical referents. However, this changing attribute is not *directly* conceptualized by autonomy codes. For this one can draw on the LCT dimension of Specialization to conceptualize relations with empirical referents in terms of *epistemic relations* and to reveal that integration of 'mathematics' into 'science' involves processes of *ontic condensation* (Maton 2014: 175–84; Wolff 2017).

Nonetheless, *autonomy codes* offer a valuable start for bringing knowledge practices into view and *autonomy tours* may represent a key to pedagogically integrating mathematics into science learning. Using the notion of *targets* to enact these concepts resolves a major obstacle to generating a general model of pedagogic integration: the problem of defining 'mathematics' and 'science' when the constitutive features of each subject, and relations between them, vary across contexts. 'Targeting' these constitutive features in *translation devices* addresses the contextual and changing nature of whether practices are constructed as 'mathematics' or as 'science'. 'Targeting' also allows studies to translate the specificities of each empirical context into concepts capable of generating a general model. We can examine, for example, the role of 'autonomy tours' in integration, howsoever 'science' and 'mathematics' are defined. By targeting science in this way, we can indeed hit a bullet with a smaller bullet, while blindfolded and riding a horse.

Notes

1 Research into science education is divided into 'science education research' for schooling and disciplinary specialisms (such as 'physics education research') at university level. In our empirical examples from secondary schooling, 'science' is taught, but our argument is not limited to one discipline or level of education.
2 Work discussing 'disciplinary discourse' (e.g. Airey and Linder 2009) points towards knowledge but reduces this 'discourse' to representations of more fundamental 'ways of knowing', leading again to studies of student 'fluency' in ways of knowing – subjectivism returns.
3 In LCT this is to say that *production fields*, *recontextualization fields*, and *reproduction fields* have their own distinctive logics (Maton 2014: 47–52).
4 We must emphasize 'complemented': to replace studies of ways of knowing with analysis of forms of knowledge would be to continue taking part of the picture for the whole. Both knowledge and knowing are significant.
5 This study was funded by the Australian Research Council (DP130100481) and led by Karl Maton, J. R. Martin, Len Unsworth and Sarah K. Howard.
6 These examples were introduced in Maton and Howard (2018) and more extensively analysed here.
7 In Chapter 4 of this volume we focus on how teachers integrate a multimedia object into a specific task, so each teacher's *target* is the lesson and their *core target* is the specific task.

8 During data collection the curriculum authority was the New South Wales Board of Studies. Though renamed the New South Wales Education Standards Authority (NESA), its syllabus remains the same at the time of publication. All 'NESA' quotes are from https://syllabus.nesa.nsw.edu.au/stage-4-content/.

9 All quotes in this paragraph are from NESA.

10 In Figure 2.3 the shift is to just inside the exotic code because content and purpose concern educational knowledge or *associated non-targets* (PA–, RA–).

11 All quotes in this paragraph are from NESA.

12 Both content and purpose may be non-target but still concern educational knowledge, so embody an *associated* exotic code.

13 The pathway in Figure 2.5 shifts to the far right, indicating extremely strong relational autonomy, because creating a graph for the experiment's results is within the teacher's *inner-core target* purpose.

References

Airey, J. and Linder, C. (2009) 'A disciplinary discourse perspective on university science learning: Achieving fluency in a critical constellation of modes', *Journal of Research in Science Teaching*, 46(1): 27–49.

Bing, T. J. and Redish, E. F. (2007) 'The cognitive blending of mathematics and physics knowledge', *American Institute of Physics Conference Proceedings*, 883(26): 26–29.

Bing, T. J. and Redish, E. F. (2009) 'Analyzing problem solving using math in physics: Epistemological framing via warrants', *Physical Review Special Topics – Physics Education Research*, 5(020108): 1–15.

Blackie, M. (2014) 'Creating semantic waves: Using Legitimation Code Theory as a tool to aid the teaching of chemistry', *Chemistry Education Research and Practice*, 15: 462–69.

Conana, H., Marshall, D. and Case, J.M. (2016) 'Exploring pedagogical possibilities for transformative approaches to academic literacies in undergraduate physics', *Critical Studies in Teaching and Learning*, 4(2): 28–44.

di Sessa, A. (1993) 'Toward an epistemology of physics', *Cognition and Instruction*, 10(2/3): 105–225.

Doran, Y. J. (2018) *The Discourse of Physics*, London: Routledge.

Fredlund, T., Airey, J. and Linder, C. (2012) 'Exploring the role of physics representations: An illustrative example from students sharing knowledge about refraction', *European Journal of Physics*, 33: 657–66.

Georgiou, H., Maton, K. and Sharma, M. (2014) 'Recovering knowledge for science education research: Exploring the 'Icarus effect' in student work', *Canadian Journal of Science, Mathematics, and Technology Education*, 14(3): 252–68.

Hammer, D. (2000) 'Student resources for learning introductory physics', *American Journal of Physics*, 68(7 PER Suppl.): S52–S59.

Hammer, D. and Elby, A. (2002) 'On the form of a personal epistemology', in B. K. Hofer and P. R. Pintrich (eds) *Personal Epistemology*, Mahwah, NJ: Erlbaum, 169–90.

Lemke, J. (1998). 'Multiplying meaning: Visual and verbal semiotics in scientific text', in J. R. Martin and R. Veel, *Reading Science*, London: Routledge, 87–113.

Maton, K. (2014) *Knowledge and Knowers: Towards a Realist Sociology of Education*, London: Routledge.

Maton, K. (2016) 'Starting points: Resources and architectural glossary', in K. Maton, S. Hood and S. Shay (eds) *Knowledge-Building*, London: Routledge, 233–43.

Maton, K. and Chen, R. T.-H. (2016) 'LCT in qualitative research: Creating a translation device for studying constructivist pedagogy', in K. Maton, S. Hood and S. Shay (eds) *Knowledge-Building*, London: Routledge, 27–48.

Maton, K. and Howard, S. K. (2018) 'Taking autonomy tours: A key to integrative knowledge-building', *LCT Centre Occasional Paper 1* (June): 1–35.

Meli, K., Zacharos, K. and Koliopoulos, D. (2016) 'The integration of mathematics in physics problem solving: A case study of Greek upper secondary school students', *Canadian Journal of Science, Mathematics and Technology Education*, 16(1): 48–63.

O'Halloran, K. (2010) 'Mathematical and scientific forms of knowledge: A systemic functional multimodal grammatical approach', in F. Christie and K. Maton (eds), *Language, Knowledge and Pedagogy*, London: Continuum, 205–36.

Parodi, G. (2012) 'University genres and multisemiotic features: Accessing specialized knowledge through disciplinarity', *Forum Linguistico*, 9:4: 259–82.

Planinic, M., Milin-Sipus, Z., Katic, H., Susac, A. and Ivanjek, L. (2012) 'Comparison of student understanding of line graph slope in physics and mathematics', *International Journal of Science and Mathematics Education*, 10: 1393–414.

Quinnell, R., Thompson, R. and LeBard, R.J. (2013) 'It's not maths; it's science: Exploring thinking dispositions, learning thresholds and mindfulness in science learning', *International Journal of Mathematical Education in Science and Technology*, 44(6): 808–16.

Redish, E. F. (2014) 'Oersted lecture: How should we think about how our students think?', *American Journal of Physics, 82*: 537–51.

Redish, E. F. (2017) 'Analysing the competency of mathematical modelling in physics', in, T. Greczyło and E. Debowska (eds) *Key Competences in Physics Teaching and Learning*, Switzerland: Springer, 25–40.

Redish, E. and Kuo, E. (2015) 'Language of physics, language of math: Disciplinary culture and dynamic epistemology', *Science and Education, 24*: 561–90.

von Glasersfeld, E. (1995) *Radical Constructivism*, Washington, DC: Falmer.

Wolff, K. (2017) 'A language for the analysis of disciplinary boundary crossing: Insights from engineering problem-solving practice', *Teaching in Higher Education*, 23(1): 104–19.

3

CONSTELLATING SCIENCE

How relations among ideas help build knowledge

Karl Maton and Y. J. Doran

Comprising a simple explanation of a complex phenomenon,
namely complex explanations of seemingly simple phenomena.

Introduction

Science is complex. This is not simply a matter of the extraordinarily large number of ideas, practices and beliefs that comprise science; it also concerns the manifold ways in which they are related together. These complex relations pose challenges for teaching science. Even after isolating a topic and deciding the level of detail to teach, questions arise of where to begin and how to proceed through the diverse connections among ideas that constitute a scientific explanation. To address these questions requires an understanding of different kinds of relations among ideas. Yet, there remains a need for concepts that reveal and analyze those relations, thanks to knowledge-blindness and atomism.

'Knowledge-blindness' (Maton 2014) describes the way the forms taken by knowledge remain unseen by research. Most approaches instead explore either kinds of knowers or ways of knowing. Sociologically-inflected approaches emphasize that 'knowledge' reflects dominant interests and focus on *whose* knowledge is taught and learned (Ellery 2017), a concern with kinds of knowers that is increasingly salient in calls to 'decolonize' science (Adendorff and Blackie 2020). More common in science education are psychologically-inflected approaches that assume 'knowledge' comprises mental processes and so focus on 'conceptions' and ways of thinking (see Georgiou *et al.* 2014). For example, the notion of 'threshold concepts' appears to highlight relations among ideas: they are 'concepts that bind a subject together' (Land *et al.* 2005: 54). However, the nature of those relations is not analysed. A 'threshold concept' is defined as 'opening up a new and previously inaccessible way of thinking about something' (Meyer and Land 2003: 1). It is identified by being 'transformative', 'integrative', 'irreversible', and often 'troublesome' for students and distinguished from a 'core concept' that 'progresses understanding of the subject' without transforming students' ways of thinking (Meyer and Land 2003: 4). Leaving aside their ill-defined nature (Salwén 2019), these concepts focus entirely on ways of knowing. The forms taken by 'core' and 'threshold' concepts are not part of the picture. Taken together, that research is concerned with kinds of knowers or ways of knowing means that knowledge as an object of study in its own right, one that takes different forms, forms which have effects for all kinds of issues, is left out of the picture.

The second reason, atomism, describes how knowledges (and ways of knowing) are viewed atomistically, as if theories, explanations, etc. are simply collections of individual ideas. For example, diSessa (1993) defines 'phenomenological primitives' or 'p-prims' as knowledge in physics that students intuitively believe to be irreducible features of reality. As well as focusing on knowing, this notion illustrates atomism: how these ideas relate to other ideas to form an explanation remains unclear. As diSessa states, the approach views 'knowledge in pieces' (1993: 111). This is akin to listing ingredients but not describing the recipe, as if how ingredients are combined does not affect what is created. Such atomism also characterizes typologies – such as Shulman's 'PCK', 'TPCK', and Bloom's taxonomy – which list different kinds of 'knowledge' but do not conceptualize relations among those types.

Given these tendencies to knowledge-blindness and atomism, there remains a need for concepts to grasp different relations among ideas. This issue was helpfully opened up by systemic functional linguists exploring explanations in science (e.g. Rose *et al.* 1992, Unsworth 1997). Such work showed that ideas in explanations are connected in language through a range of relations of condition, cause and time. However, the issue of how relations among ideas affect how they are explained remains unclear. In this chapter, we draw on *constellations* from Legitimation Code Theory (Maton 2014) to conceptualize and visualize relations among ideas. We focus on analyzing explanations, a key genre in science education (Unsworth 1997). Our aim is to illustrate how constellation analysis can show the significance of relations among ideas for science teaching.

We begin by introducing the notion of *constellations*. We then make several simple conceptual distinctions relevant to our analyses in this chapter. These concepts are enacted in analyses of scientific explanations of tides and seasons, focusing on Year 7 of secondary school science in New South Wales, Australia. Analyzing textbooks created for this curriculum, we examine the logical relations among ideas in each explanation. Then, we examine how each explanation is taught by the same teacher in the same unit of study. From these analyses we argue that relations among ideas in the logic of explanations may affect how those explanations are taught in a school classroom. We conclude by considering how constellation analysis can provide educators with a pedagogic tool for making visible relations among the ideas they are teaching, and researchers with an analytic tool for making visible how knowledge changes when recontextualized between research, curriculum and classroom practice.

Constellations

A way of grasping relations among ideas is provided by *constellation analysis*, which forms part of *cosmological analysis* from Legitimation Code Theory (LCT). Cosmological analysis describes any set of stances (e.g. ideas, beliefs, practices, etc.) as a selection from a larger set of possible stances that has been arranged into a particular pattern or *constellation*, *condensed* with meanings, and *charged* with valuations, according to a particular *cosmology* or worldview (Maton 2014: 148–70).

Analogy from astronomy

As a way into the approach, consider the notion of 'constellations' in astronomy. On a clear night without light pollution one can see an enormous number of stars. As an example, the left image in Figure 3.1 shows some of the stars visible in a small part of the sky in the northern hemisphere. Of those stars, a small number have been selected and arranged into a pattern that is the *constellation* of 'Taurus' (middle in Figure 3.1). There may also be smaller *clusters*; for example, 'Pleiades' is a cluster of stars that lie within Taurus. As well as being constellated, the stars are *condensed* with meaning. For example, since the ancient Mesopotamians, Taurus has been associated with the image of a bull (right in Figure 3.1). These meanings are *charged* positively, neutrally or negatively, to varying degrees. For example, in modern astrology Taurus signals such attributes as 'creativity', 'affectionate', and 'grasping'.

The selection, arrangement, meaning and valuation of constellations may vary across place and over time. For example, some stars are visible only from specific locations, but this may change – 'The Southern Cross' was visible in the northern hemisphere before the fifth century. The meanings of a constellation may also change. For example: the Zuni people of New Mexico call the Pleiades cluster 'seed stars', as their position was traditionally used to determine the time of the year to plant seeds; and Pleiades is the logo of the car manufacturer Subaru, whose advertising attempts to associate the symbol with such notions as 'reliability'. The constellations themselves also vary, reflecting differences in *cosmology* or worldview. For example, Figure 3.2 shows Aldebaran, a star in Taurus, located within constellations from (left to right) Inuit, Korean and Maori cultures.[1]

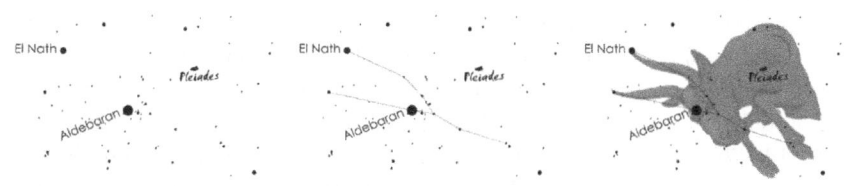

FIGURE 3.1 Taurus – some stars before constellating; stars constellated into Taurus; overlaid with image of bull (Northern hemisphere view)

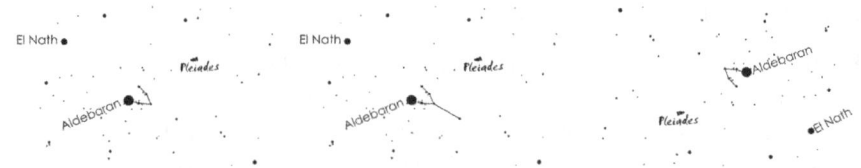

FIGURE 3.2 Aldebaran in constellations from Inuit, Korean and Maori cultures (hemisphere view is that of each culture)

LCT concepts

There are limits to any analogy, but the astronomy example offers a way into understanding the LCT approach to analyzing practices.[2] Cosmological analysis is centred on the five words italicized above:

- *clusters* are groupings of nodes (such as ideas, beliefs or practices);
- *constellations* are larger grouping of nodes that may include clusters;
- *condensation* is how nodes, clusters and constellations are imbued with meanings;
- *charging* describes the valuations given to nodes, clusters and constellations; and
- *cosmology* refers to the organizing principles underlying the selection, arrangement and valuation of nodes in a constellation, which are revealed by analyzing their *legitimation codes*.

Reflecting its sociological nature, LCT holds that a constellation has coherence from a particular point in social space and time, to actors with a particular cosmology, and that cosmologies (and so constellations) are subject to contestation, vary across contexts, and change over time (Maton 2014: 148–70).

This description is intentionally abstract, to reflect the wide applicability of cosmological analysis. The constellation being analyzed may be a scientific theory, religion, political system, ideology, sport, dance, song, machine, etc. – its nodes may be ideas, beliefs, institutions, body movements, sounds, machine parts, etc. Similarly, analysis may focus on many different issues about these constellations, using different cosmological concepts. Here we shall explore relations among stances using only the concepts of *clusters* and *constellations*. We shall not explore the nature of the meanings being related (using *condensation* and *charging*) nor reveal the underlying principles generating, maintaining and changing stances (*cosmologies*). Our focus is a *constellation analysis* of how stances are related together, rather than a *cosmological analysis* of the organizing principles underlying those relations.[3]

Constellation analysis can itself be used in different ways. For example, Maton (2014: 148–70) illustrates a synchronic analysis of ideas. Analyzing claims by advocates of constructivism, Maton shows they construct two constellations of stances on curriculum, pedagogy, assessment and many other issues, that are given such labels as 'student-centred' and 'teacher-centred'. Relations *within* each constellation are constructed as essential: choosing one stance is viewed as choosing all other stances in the same constellation. Relations *between* the constellations are constructed as oppositional: one cannot select stances from both. The two constellations are also portrayed as the only options available. Maton argues that these relations have effects, such as narrowing what is viewed as possible in education by excluding stances either outside or spread between the constellations. Similarly, Glenn (2016) analyzes beliefs about climate change, showing how different groups of people constellate different ideas together and charge those ideas differently, in ways which mean they are more or less open to scientific evidence for climate change.

Another way of using constellation analysis is to explore how stances are selected, linked and given meaning over time to build practices. For example, Lambrinos (2020) reveals how ballet teachers bring together sets of behaviours, dispositions and movements to teach both dance and how to be a dancer. Such analysis can show how clusters or even whole constellations are condensed into a new node, which can then be constellated with more nodes. Figure 3.3, for example, illustrates how a ballet teacher teaches an exercise called 'springs' by linking the instruction 'jump' with other instructions ('sink', 'feet in 1st position', etc.), where 'jump' (solid black in Figure 3.3) has itself been condensed with meanings ('powerful', 'straight legs', etc.). In this way, Lambrinos shows how words, gestures and movements are brought together to create complex constellations. Other studies focus on the building of values in texts. For example, Tilakaratna and Szenes (2020) show how successful student 'reflective writing' assignments cluster and condense meanings to align with disciplinary values, and Doran (2020) reveals the rhetorical strategies of a highly influential text that cluster, condense and charge values to effect change within an intellectual field.

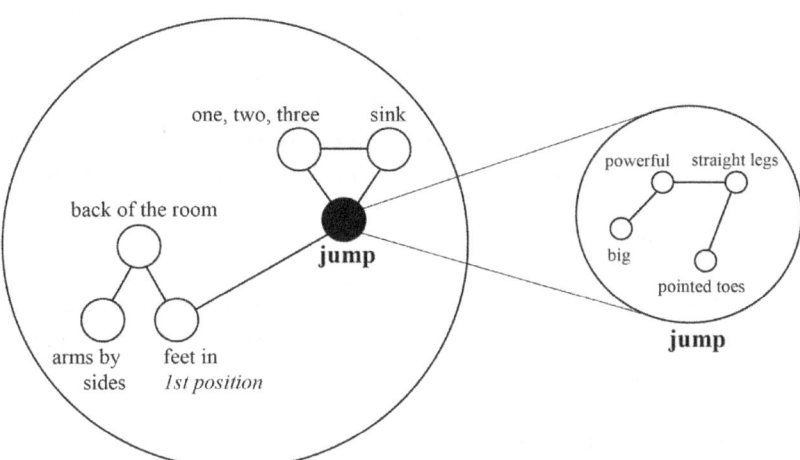

FIGURE 3.3 Example of constellating and condensing movements in ballet teaching (Lambrinos 2020: 91)

However, constellation analysis is still in its infancy. One area for development is identifying different *kinds* of relations within a constellation. Thus far, studies have mostly focused on differences among constellations in terms of their cosmologies or underlying legitimation codes (Maton 2014: 148–70). As yet few studies explore different relations *within* a constellation.[4] In this chapter we contribute to exploring this issue.

Three simple distinctions

To illustrate the significance of different relations among ideas, we shall make three simple distinctions relevant to our analyses in this chapter, between 'independent' and 'dependent' links, 'base' and 'supplementary' clusters, and 'assembling' and 'aggregating' when building constellations.

First, we distinguish between:

- *independent links,* where the meanings of a node or cluster are independent of other nodes or clusters in the constellation; and
- *dependent links,* where the meanings of a node or cluster depend on relations to other nodes or clusters in the constellation.

For example, in the explanation of tides (below), the node 'The Earth is divided into water and solid components' does not depend on other nodes in the explanation, generating independent links with those nodes. It does not depend, for example, on the idea that 'The Moon has gravity, which is stronger the closer things are to the Moon'. This is not to suggest that the ideas expressed by such nodes are always independent; they will be related to ideas in *other* constellations, such as explanations of the Earth, planets, etc. However, this highlights where nodes create independent links with other nodes *within the constellation in question*. In contrast, in the tides explanation a node that describes the different gravitational pulls of the Moon on the water and solid components of the Earth links to two other nodes in the explanation – 'The Earth is divided into water and solid components' and 'The Moon has gravity, which is stronger the closer things are to the Moon' – by describing their implications when combined. It thereby generates dependent links with those two nodes within the constellation.[5]

Second, we distinguish between:

- a *base cluster* or set of nodes that creates a basic version of the constellation; and
- *supplementary clusters* or sets of nodes that serve to elaborate, augment and refine the base cluster.

For example, in the explanations of tides, the textbooks and teaching we analyze draw on four nodes to create a basic explanation of tides (the base cluster), and elaborate on this explanation by employing other clusters of nodes to explain daily variation in tides, and 'spring' and 'neap' tides (two supplementary clusters). Together all three clusters constitute the explanation, but one cluster offers a 'base' on which the other clusters elaborate. This is *not* a distinction between fundamental and peripheral, essential and inessential nodes, but rather highlights two roles played by clusters of nodes in the explanations we analyze in this chapter.

Third, we distinguish between two ways that constellations unfold over time:

- *assembling*, where nodes and clusters develop in a linear and incremental fashion; and
- *aggregating*, in which nodes or clusters are developed separately, in a multilinear fashion.

In *assembling* the constellation is likely to grow from one origin; in *aggregating* the constellation involves multiple separate parts, each of which may be assembled on its own before being combined.[6] As we shall illustrate, the tides explanation appears to lend itself to assembling nodes and clusters in a linear manner, while the seasons explanation has a range of potential ways of aggregating nodes and clusters together.

These three distinctions are not the only relations within constellations created by scientific explanations, let alone within constellations generally. Our aim is not to conceptualize all relations among ideas but to make intentionally simple distinctions that demonstrate a simple point: different relations among ideas matter. Conversely, not all constellations involve these distinctions: they may comprise only dependent or independent nodes, have no base or supplementary clusters, and remain static. The distinctions are thus not generic characteristics – they reflect our specific focus and aim.

Analyzing two explanations: Of tides and seasons

Our focus is on analyzing two scientific explanations. Our aim is to show that, though they appear similar, the explanations involve different relations among ideas that may be significant for how they are taught. Both are core topics in Year 7 of secondary school science in New South Wales (NSW), Australia. Both concern widely-known phenomena in the natural world: the tides and the seasons on Earth. Both phenomena appear *prima facie* simple: water reaches higher and lower through the day (tides) and the temperature goes higher and lower over the year (seasons).[7] However, we shall show that explanations of tides and seasons involve complex constellations of ideas brought together in distinctive ways.

We analyze each explanation in two ways. First, we create *schematic constellations* of the logic of each explanation according to textbooks aimed at Year 7 secondary school science in NSW. This is not analyzing how textbooks sequence their explanations or build constellations; it does not describe a specific textual or pedagogic expression. Rather, a 'schematic constellation' is a synchronic representation of key nodes and how they are logically related – a snapshot of *the logic of the explanation*. To create the constellation, we identify nodes and links that the textbooks contain or assume in order to make sense.[8] The resulting constellation diagrams are akin to transit maps that show not specific journeys but rather the roads, stations, connections, routes, etc. They are a composite from analyses of textbooks known as: Oxford Insight (Zhang *et al.* 2013), Nelson iScience (Bishop *et al.* 2011), Pearson (Rickard 2011), Core Science (Arena *et al.* 2009), and Science World (Stannard and Williamson 1995).

Second, we create *pedagogic constellations* showing how each explanation is taught by the same teacher during the same unit of secondary school science.[9] Our focus is to explore whether the logic might shape the pedagogic, that is whether relations among ideas in the logic of an explanation might affect how that explanation is taught. To return to the metaphor, these pedagogic constellations describe specific journeys across the terrain shown by the schematic constellations.

This analysis does not, of course, offer a comprehensive account of how tides and seasons are explained in schooling. As stated above, constellations often vary across time and space. In other contexts, different explanations may include more or fewer nodes and different links. The analysis is thus limited to our empirical examples. Moreover, our examples are not intended to demonstrate best practice and whether the explanations are accurate or complete is not our concern. Our aim is simply to explore how relations among ideas are expressed and their potential significance.

In summary, we shall argue that the schematic constellation for tides is less complex than that for seasons. One effect of this difference is, we suggest, shown by pedagogic constellations: the tides explanation lends itself to an *assembling* form of teaching that proceeds in a linear fashion through successive clusters, while the seasons explanation lends itself to an *aggregating* form of teaching that proceeds in a more patchwork fashion. We conjecture that the forms taken by relations among ideas in the logic of an explanation (schematic constellation) may shape how the explanation is likely to be taught (pedagogic constellation). To reach these conjectures, we analyze tides and of seasons, in turn.

Explaining tides

Schematic constellation of textbooks

From analyses of sections on 'tides' in the five textbooks, we developed a composite of key nodes in their explanations. We begin by summarizing these nodes as simply as possible, in our own words. In **bold** are words included in the diagrams that follow.

A. The Earth is divided into **water and solid** components.
B. The **Moon** has **gravity**, which is stronger the closer things are to the Moon.
C. Together node A (water and solid) and node B (Moon's gravity) mean that the **Moon's** gravitational **pull** is strongest for the water on the Earth closest to the Moon, less strong for the Earth's solid, and weakest for the water on the Earth furthest from the Moon.
D. Node C produces bulges of water on the parts of the Earth that are closest and furthest away from the Moon, which we experience as '**high** tides', and no

bulges on the parts of the Earth that are neither closest nor furthest, which we experience as '**low** tides'.

E.   The **Earth rotates**.

F.   Earth's rotation (node E) combined with the bulges of water created by the Moon's gravity (node D) leads to the experience of **daily variation** of tides as the Earth moves through the bulges.

G.   The **Moon orbits** the Earth.

H.   The **Sun's gravity** pulls on the water and solid components of the Earth.

I.   Combining the Moon's orbit of the Earth (node G) and the Sun's gravitational pull (node H) with the daily variation of tides (node F) leads to variation in the size of tides. When the Sun and Moon line up, the tides vary the most (highest highs and lowest lows), which is known as the '**spring** tide'; and when the Sun and Moon are perpendicular, the tides vary the least (with the lowest highs and highest lows), which is known as the '**neap** tide'.

Figure 3.4 represents this description as a constellation diagram. Nodes which generate *independent links* are in squares and nodes which generate *dependent links* are in

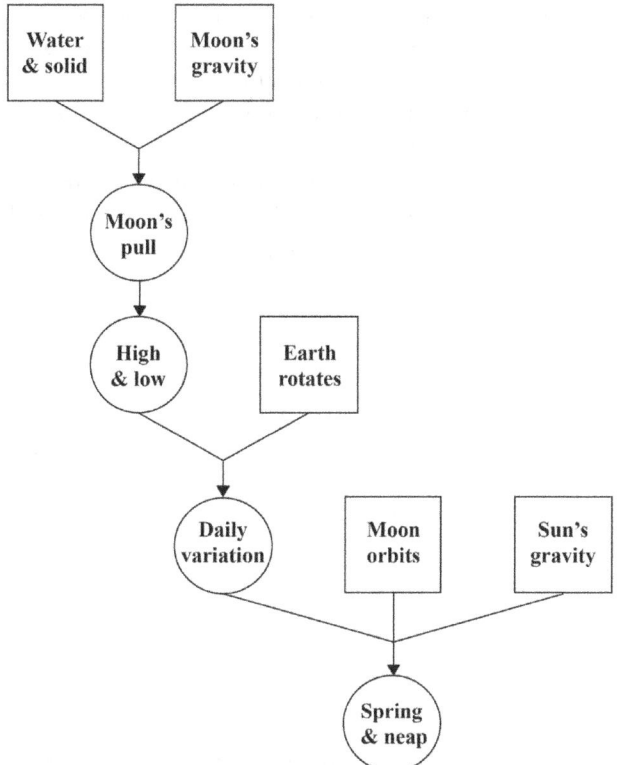

FIGURE 3.4   Schematic constellation of the tides explanations across textbooks

circles. Dependent links are shown by lines with arrows that indicate the direction of implication.

Links, clusters and form

Though 'tides' may seem a simple empirical phenomenon, its explanation is relatively complex, involving different links and clusters. In terms of links, the schematic constellation is characterized by both independent and dependent relations among nodes. As Figure 3.4 shows, the schematic explanation begins with two nodes that create independent links: the Earth's components of **water and solid** and the **Moon's gravity** are not dependent on other nodes *in this explanation*. In contrast, how the **Moon's pull** affects Earth's components comprises implications of bringing together those two nodes, generating dependent links (shown by the arrow). The node introducing **high and low** tides outlines the implications of the Moon's pull for creating bulges of water. This reaches the notion of 'tides'. That the **Earth rotates** is independent. The **daily variation** of tides creates a dependent link by describing implications of bringing together the **Earth's rotation** with **high and low** tides. That the **Moon orbits** the Earth and the **Sun's gravity** are both independent. To reach the notions of **spring and neap** tides, these two nodes are combined with **daily variation**, creating dependent links with all three nodes. In summary, a series of independent nodes establish factors or phenomena and dependent nodes relate those phenomena together.

In terms of clusters, the schematic constellation for the textbooks' explanation of tides comprises a *base cluster* and two *supplementary clusters*. As shown in Figure 3.5a, the top four nodes form a base cluster that creates a basic explanation of tides. The two supplementary clusters are *successive* in that each progresses from a logically preceding node. As shown by Figure 3.5b, the cause of **high and low** tides is the starting point

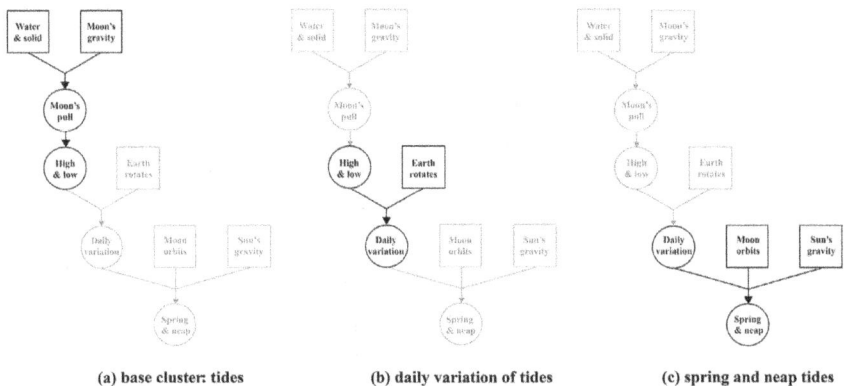

(a) base cluster: tides   (b) daily variation of tides   (c) spring and neap tides

FIGURE 3.5 Schematic constellation for tides – base and supplementary clusters

for a supplementary cluster with the **Earth's rotation** that augments the basic explanation to reach **daily variation** in tides. As shown by Figure 3.5c, **daily variation** is part of a second supplementary cluster with the **Moon orbits** and **Sun's gravity** that elaborates the preceding explanation to describe **spring and neap** tides.

In terms of its form, the schematic constellation does *not* show how the explanation unfolds over time. However, it does reveal that its logical links are relatively linear: the simple explanation of the base cluster is augmented through successive supplementary clusters. There are no parallel clusters elaborated or large numbers of separate nodes described which then require integration. The implications of this relative simplicity will, however, only become apparent in contrast with our analysis of 'the seasons', further below.

A pedagogic constellation for tides

We now turn to how an explanation of tides was taught in a secondary science classroom in comparison to this schematic constellation. Our example is part of a lesson within a unit on 'Earth and space sciences' from Year 7 of secondary school, NSW, that explores such topics as 'day and night', 'the seasons' and 'tides'. We focus on a lesson phase in which the topic of 'tides' is introduced by the teacher and a video is played that details an explanation.

Introducing the topic

Immediately prior to addressing 'tides', the teacher shows a video about the Moon and the Earth. She then asks the class what might cause tides. A student suggests 'the Moon's gravitational pull', which was mentioned in the video, and the teacher elaborates:

> Yes, very good. So, because the Moon has gravity it actually pulls the mass of water towards itself. Wherever the Moon is, that's where the high tide will be and wherever the Moon isn't, that's where the low tide will be.

So, the teacher links the **Moon's gravity** and the Earth's **water and solid** composition by how the **Moon's pull** affects the water and thence to a simple description of **high and low** tides. She thereby provides a succinct, Moon-focused equivalent to the base cluster outlined above (Figure 3.5a) before starting the video on tides. As this highlights, pedagogic constellations may form part of a series in which teachers and student build on preceding discussions or look ahead to future topics. They venture onto the terrain shown by the transit map from where they have just been, in this case discussing the Moon.

A video explanation

The teacher plays a video entitled 'Watching the tides' in which an astronomer explains the causes of tides.[10] The video begins by summarizing key factors:

> Tides are caused by the gravitational pull of the Moon and the Sun and the rotation of the Earth. The Earth is not a solid sphere like we think. It's actually kind of squishy. Especially this layer of water we have on the outside – the oceans.

This is accompanied by an animation showing the Moon orbiting the Earth. It thus begins by highlighting: the **Moon's gravity**, the **Sun's gravity**, the **Earth's rotation**, the Earth's composition as **water and solid**, and the **Moon's orbit**. Thus, as shown by Figure 3.6a, the video begins by introducing the five independent nodes.[11]

The video then begins creating dependent nodes that link these independent nodes together:

> So the gravity of the Sun and the Moon actually kind of squeeze or stretch the Earth and its oceans out into a couple of bulges. One under the Moon, one on the other side of the Earth.

This links 'the Earth and its oceans' (**water and solid**) with the **Sun's gravity** and the **Moon's gravity** to explain that the oceans are pulled into bulges, while the accompanying animation shows bulges on the parts of the Earth closest to and furthest from the Moon, creating a simple version of the node stating that the **Moon pulls** on Earth's components differently. The video continues:

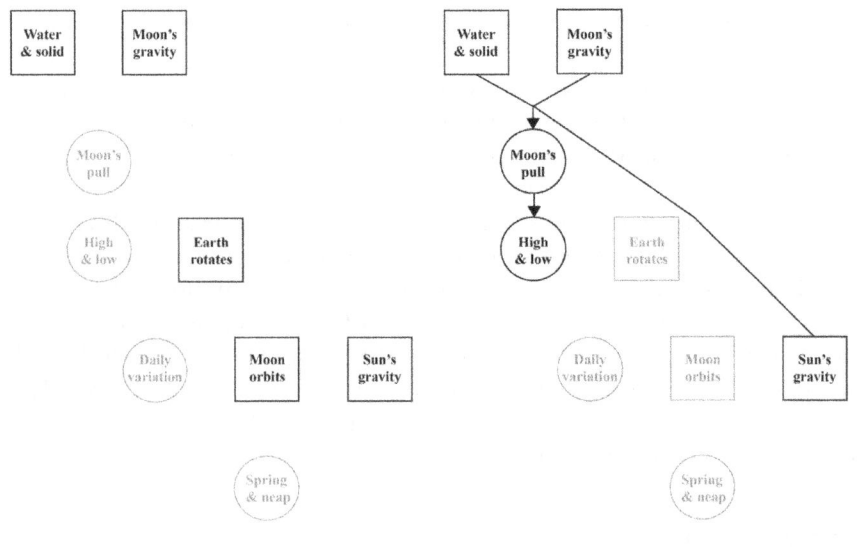

(a) setting out independent nodes   (b) a base cluster

FIGURE 3.6 Pedagogic constellation for tides – independents and base cluster

And as the Earth rotates over the course of the day, you, standing on the surface of the Earth, move along with the Earth's surface into these bulges. And we experience that as the rising and lowering tides.

This is a simple version of **high and low** tides. As illustrated by Figure 3.6b, the video has now outlined a base cluster echoing that of the schematic constellation. There are two main differences. First, the **Sun's gravity** has been recruited verbally into the base cluster, though the animation does not include the Sun and its role is not discussed. Second, the key attribute of the **Moon's gravity** – that it is stronger the closer things are – has not yet been mentioned. This issue is, though, immediately discussed (emphasis added):

> Now it's easy to see how on the side facing the Moon or the Sun you can get this bulge of ocean. You can imagine the gravity pulling the oceans up into a bob or a bubble. But it's not as easy to understand why there's a bulge on the other side as well. And the easiest way to describe that is: *the Moon's gravity is stronger, of course, the closer you get to it*. So, on the side of the Earth close to the Moon, the Moon has a stronger pull. So while the oceans on the Moon side get pulled more strongly than the general Earth does, on the other side it's kinda opposite. The pull on the oceans on the far side are less than the pull on the Earth. So that far bulge actually gets created … think of it as the Earth being pulled out from under the oceans, a little bit.

The video thereby completes the base cluster (Figure 3.6b). The attribute of the **Moon's gravity** (in italics) has been added and brought together with the Earth's composition as **water and solid** to describes implications for the **Moon's pull** on those components, and so the creation of bulges of water or **high and low** tides.

In short, the video first creates a preliminary, intuitive version of the base cluster and then uses the counter-intuitive nature of the Moon's gravity creating a bulge on the *far* side of the Earth as a way of completing that basic explanation. Next, the video links **high and low** tides with the **Earth's rotation** to describe implications for **daily variation** of tides:

> You get two high tides a day because as the Earth rotates, we rotate through these two bulges.

As shown in Figure 3.7c, this echoes the first supplementary cluster of the schematic constellation. The video then creates the second supplementary cluster – Figure 3.7d. The role of the **Sun's gravity** is shown in the accompanying animation and linked with the **Moon's orbit** to introduce '**spring** tide' **and** '**neap** tide':

> Both the Moon and the Sun play a part in tides. Each one pulls. And when the Sun and the Moon combine their forces – that is, when they're both acting together – we get much stronger tides than usual. Higher highs and lower

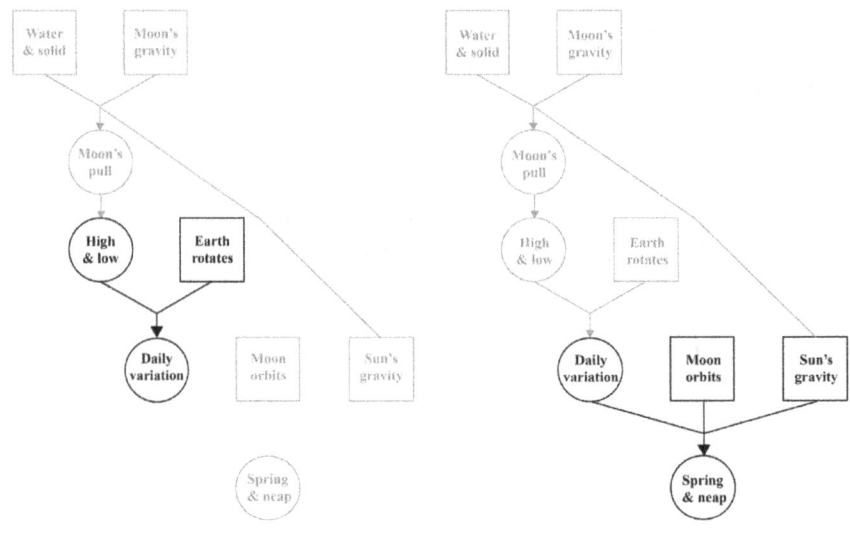

(c) first supplementary cluster   **(d) second supplementary cluster**

FIGURE 3.7 Pedagogic constellation for tides – supplementary clusters

lows – we call these 'spring tides'. The name 'spring tide' doesn't have anything to do with the season of spring, but we get them about twice a month at new Moon and at full Moon, when the Earth, the Moon and the Sun are all lined up and the gravity of the Sun and the Moon are acting together. 'Neap tide' is when the tidal effects of Sun and Moon are kind of cancelling each other out or making each other not as extreme. And that happens around first and third quarter phases of the Moon. The Sun is in one part of the sky and the Moon is ninety degrees around and they're kind of pulling in different directions. So you get lower highs and higher lows during the neap tide.

This is the end of the video's explanation of tides, completing the constellation.

The pedagogic constellation thereby contains the same nodes, links and clusters as the schematic constellation. The form of constellation building here is what we defined as *assembling*. After introducing the independent nodes, the video selected and brought together, through dependent nodes, a subset to create a base cluster. It then added another independent node and drew out its implications through a dependent node, repeating this move to complete the explanation. It set this out in a linear and incremental fashion, accreting new nodes and linking them to existing nodes. A further attribute is how closely the pedagogic ordering of the nodes and clusters in the video matches the logic of the schematic constellation. There are many good reasons why teaching may differ from the logic of an explanation. For example, an educator may choose to build on what they have been discussing (as the teacher did in her introduction) or to wait before introducing an attribute in order to begin from shared experiences or intuitive common sense (as the video did here).

However, in this case the ordering remains remarkably similar. As we shall see, this contrasts with the unfolding of the explanation of seasons.

Explaining seasons

Schematic constellation of textbooks

From analyses of explanations of 'seasons' in the five textbooks, we developed a composite description of their key nodes, which we again state as simply as possible, in our own words, and with **bold** indicating nodes in constellation diagrams (starting with Figure 3.8).

A. The **Earth is tilted** on its axis at an angle of 23.4 degrees.
B. The Earth is divided into northern and southern **hemispheres**.

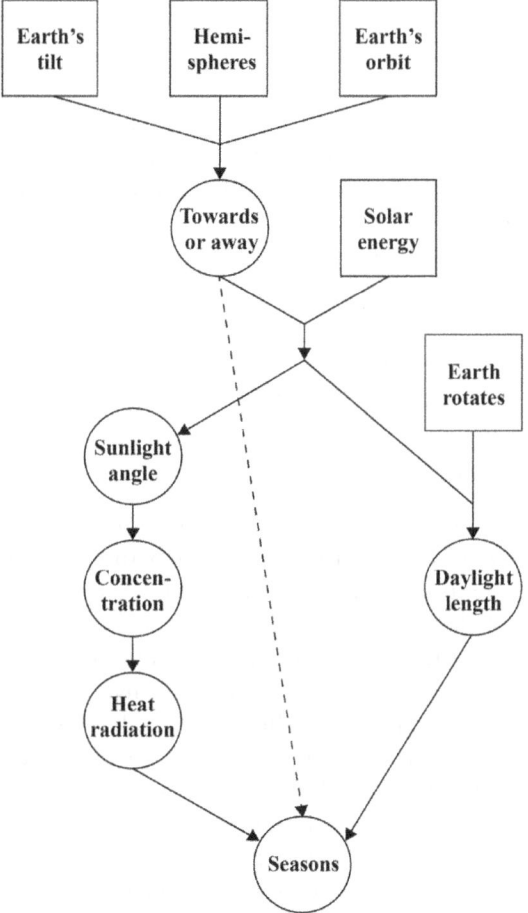

FIGURE 3.8 Schematic constellation of the seasons explanations across textbooks

C. The **Earth orbits** the Sun.
D. The Earth's tilt (node A), division into hemispheres (node B), and orbit of the Sun (node C) together mean that the Earth's northern and southern hemispheres point **towards or away** from the Sun at different times of the year.
E. The Earth receives **solar energy** from the Sun's rays.
F. That hemispheres point towards or away from the Sun (node D) and that the Earth receives solar energy from the Sun (node E) mean that **sunlight** hits each hemisphere at different **angles** through the year: when a hemisphere is pointing towards the Sun the angle of its rays are more direct; and when a hemisphere is pointing away from the Sun the angle of its rays are less direct.
G. Different angles of sunlight (node F) means that the **concentration** of light changes in each hemisphere through the year: when a hemisphere is pointing towards the Sun, the more direct sunlight it receives is concentrated in a smaller area; and when a hemisphere is pointing away from the Sun, the less direct sunlight it receives is concentrated in a larger area.
H. Differences in concentration of sunlight (node G) means that the amount of **heat radiation** in each hemisphere varies through the year.
I. The **Earth rotates**.
J. That hemispheres point towards or away from the Sun (node D), the Earth receives solar energy (node E) and the Earth rotates (node I) together mean that different parts of the Earth experience different **length of daylight** at different times of the year.
K. Variations in heat radiation (node H) and/or daylight length (node J) leads to variations in temperature in each hemisphere through the year that are called '**seasons**'. This can also be more simply put as: that hemispheres point towards or away from the Sun (node D) leads to variations in temperature in each hemisphere through the year that are called 'seasons'.

Links, clusters and form

Figure 3.8 represents the description as a schematic constellation. Like that for tides, the constellation involves nodes that generate both independent (squares) and dependent (circles) relations with other nodes. Also like tides, it exhibits a base cluster and two supplementary clusters. However, there are significant differences in how the seasons constellation relates together its constitutive ideas.

The schematic constellation begins with three independent nodes describing **Earth's tilt**, **hemispheres** and **orbit**. These nodes are brought together to describe their implications for hemispheres pointing **towards or away** from the Sun, generating dependent links with all three. From here one can proceed directly to implications for variations in temperature in hemispheres through the year, reaching **seasons** through a dependent link. As shown in Figure 3.9a, this represents a base cluster. There are then two supplementary clusters that develop this explanation. At this point the constellation becomes more complex than that of tides, in two ways.

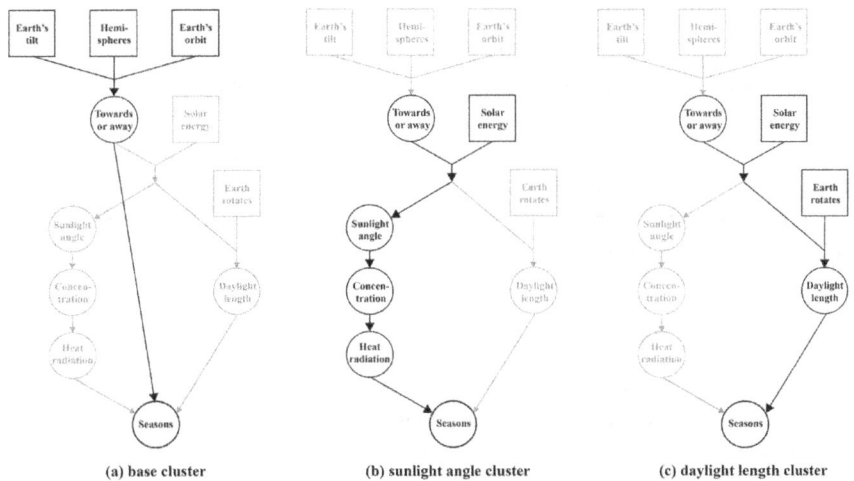

(a) base cluster   (b) sunlight angle cluster   (c) daylight length cluster

FIGURE 3.9 Schematic constellation for seasons – base and supplementary clusters

First, the supplementary clusters offer parallel routes through the logic of the explanation. What we term the 'sunlight angle cluster' recruits the nodes highlighted in Figure 3.9b. This brings together the node describing hemispheres pointing **towards or away** from the Sun with the independent node that the Earth receives **solar energy** from the Sun's rays to describe their implications for variations in the **angle of sunlight** received by hemispheres. It then elaborates implications of these variations in sunlight angle for the **concentration** of sunlight, the implications of variations in concentration for **heat radiation**, and in turn the implication of variations in heat radiation for temperatures, which reaches **seasons**. A second supplementary, which we term the 'daylight length cluster', is shown in Figure 3.9c. This too brings together the nodes on pointing **towards or away** from the Sun and **solar energy** but this time adds the independent node that the **Earth rotates** to describe the implications of these three together for variations in the **length of daylight** in hemispheres through the year. It then draws out the implications of these variations in daylight length for temperatures, reaching **seasons**. Crucially, these two clusters are not successive but parallel: each cluster reaches 'seasons' separately. The logic allows seasons to be explained through either cluster or both clusters together.

A second difference with tides is that the supplementary clusters do not elaborate the base cluster in the same way. In tides, each supplementary cluster adds to the logically preceding cluster, proceeding from 'tides' to 'daily variation in tides' to 'spring and neap tides'. In contrast, here the supplementary clusters 'unpack' the implications of hemispheres pointing **towards or away** from the sun for creating **seasons**. They are akin to focusing in on the long dashed arrow in Figure 3.9a and providing more detailed explanations of that relation. Put another way, the supplementary clusters for tides add relations to new destinations along one route (tides – daily variation – spring/neap) while the supplementary clusters for seasons clarify and deepen an existing relation by adding new routes to the same destination (seasons). These routes show the complexity latent within the link between '**towards or away**' and '**seasons**'.

In short, the seasons constellation is less linear and successive than that of tides, offering more options for navigating its logic to create an explanation. We now turn to explore how this difference might be reflected in how explanations of the seasons are taught in a classroom.

A pedagogic constellation for seasons

The example we discuss comprises a lesson on 'seasons' in the same Year 7 secondary school classroom, immediately preceding the lesson on tides analyzed above. The teacher begins the topic by showing how ripe the topic is for misunderstandings.[12] She plays a 'vox pop' video in which adults are asked for causes of seasons and suggest such mistaken beliefs as the equator, changing distance from the Sun, and Earth's elliptical orbit. The teacher states this shows that 'You have to think carefully about what we're doing'. Unlike 'tides', she thus begins by highlighting the complexity of the constellation. This complexity, we argue, is reflected in the number of times she takes the class on different routes to reach an explanation. Specifically, she takes the class through: (a) a 'daylight length' route; (b) a 'sunlight angle' route; (c) a 'base cluster' route; and (d) a composite route that includes ideas from all three clusters. These routes are discrete: they are separated by class discussions of related ideas (such as the names of longest and shortest days) or by student questions (such as whether leap years affect the Earth's orbit) that are not woven into explaining the seasons. We shall go through each explanation briefly, using the schematic constellation as a basis for comparing these routes.

(a) A 'daylight length' route
The teacher begins from where the previous lesson ended, with changes in the length of days through the year:

> In the summer, days are longer, and in the winter, days are shorter. When the days are longer, that means there is more time for the Sun to heat the Earth. So that means that the temperature is warmer. Okay?

In terms of the schematic constellation, the teacher begins with the effects outlined as '**daylight length**' and '**seasons**'. She then refers to an experiment students conducted in a recent lesson that involved heating a wooden block with a lamp to model the effect of the Sun's rays on the Earth:

> So remember that experiment we did when we put the thermometer in the block? As the time at which we kept that light on the block increased the temperature. It's the same thing that happens with days. When the days become shorter, we have overall lower temperatures. Because we have long days in

summer and short days in winter, we get higher temperatures, we get lower temperatures.

This brings in **solar energy** from the Sun. The teacher asks: 'why do we have some long days and some short days?'. A student answers 'because the Earth is **tilted**', and the teacher responds:

> Good. So it's got to do with the tilt of the Earth and depending on which part is tilted closer to the Sun will determine which will have longer and shorter days.

This is to say that depending on which of Earth's **hemispheres** is brought by **Earth's tilt** to be pointing **towards or away** from the Sun will determine the **daylight length** in that hemisphere.

Figure 3.10a, like all constellation diagrams in this analysis, shows the ideas and relations among ideas of each explanation rather than the exact sequencing of nodes in the teacher's discourse. As the Figure illustrates, the teacher offers here a simple 'daylight length' route through the explanation. Taken as a whole, it brings together **Earth's tilt** and **hemispheres** to reach **'towards or away'**, which, bringing in the

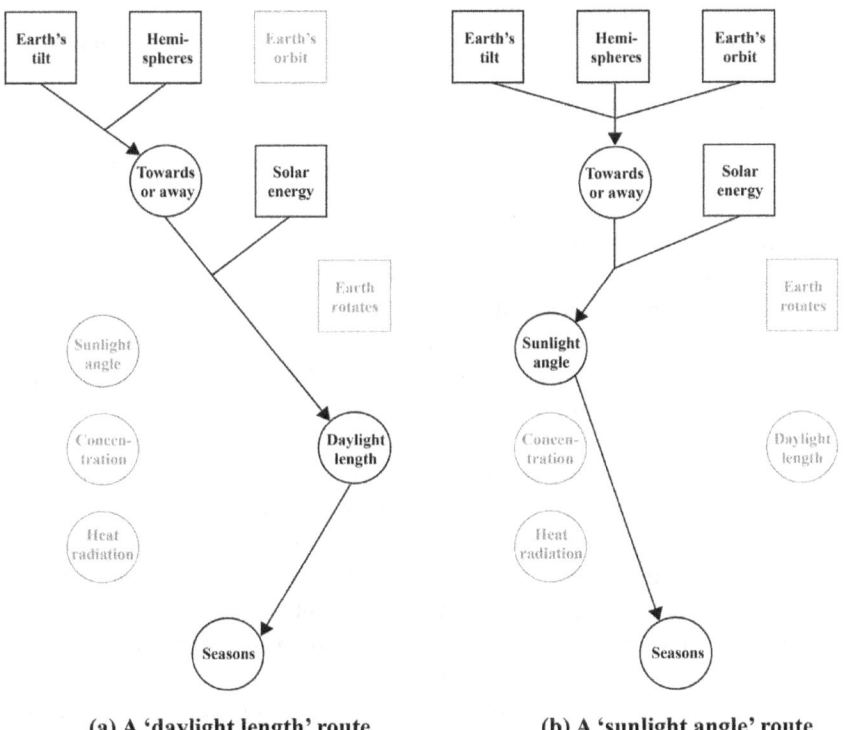

(a) A 'daylight length' route   (b) A 'sunlight angle' route

FIGURE 3.10 Pedagogic constellations for seasons – 'daylight length' and 'sunlight angle' routes

aforementioned **solar energy,** reaches **daylight length** and thence **seasons**. At this stage, the teacher did not include **Earth's orbit** and **Earth's rotation**.

(b)   *A 'sunlight angle' route*
In the next version of the explanation, the teacher shows an animation entitled 'What causes Earth's seasons', muting the sound and adding her own commentary (to which we have added the nodes discussed in brackets):[13]

> … what we're looking at here is obviously the Earth spinning on its axis [**Earth rotates**] and you can see that that axis is about 23.5 degrees from what could be the "theoretical midline" of the Earth [**Earth's tilt**]. We know that that axis holds itself. … Now, when the Sun's rays [**solar energy**] hit the Earth, the Earth has that also other "theoretical midline", the equator, that breaks it into half: northern hemisphere and southern hemisphere [**hemispheres**]. Okay? We're in the southern hemisphere and at some times the sunlight will strike, in this case, the southern hemisphere at that 90 degree angle [**sunlight angle**]. All right? And here, this is what we can sometimes call an 'oblique' … or 'glancing' [**sunlight angle**].

The first node, **Earth rotates**, is mentioned to direct students to watch the animation and not woven into the explanation. The rest – **Earth's tilt**, **solar energy**, **hemispheres** and **sunlight angle** – are introduced first as factors to be related. She then brings them together:

> Now, when this moves around the Sun [**Earth's orbit**], and holds its axis [**Earth's tilt**], because it doesn't … the axis won't change around like that, when this moves to the other side of the Sun, we will see the other side of the Earth [**towards or away**]. … now we're on the back side of the Sun. We've got spring and autumn [**seasons**]. Now we're on the opposite side. Now the northern hemisphere has summer and the southern hemisphere has winter [**seasons**]. Now, we come out to the "fourth side" of the Sun, we get autumn and spring split around [**seasons**].

The teacher then sums up this explanation with a PowerPoint slide showing the Earth (with equator and axis shown), the Sun and its rays hitting the Earth: 'So we know the Sun is hitting the Earth [**solar energy**]. We know that some parts of the Earth [**hemispheres**] will be getting the full force and some will getting those glancing rays [**sunlight angle**]' and then emphasizes that 'The thing that will change is our position in relationship to the Sun' [**towards or away**].

As illustrated by Figure 3.10b, taken as a whole this offers a 'sunlight angle' route through the explanation. In combination with the animation, the teacher shows that **Earth's tilt**, **hemispheres** and **orbit** mean that different hemispheres point **towards or away** from the Sun which, when struck by **solar energy**, means there are differences in **sunlight angle** that create **seasons**.

(c)   *A 'base cluster' route*

The third explanation comprises three activities: discussing a video animation, students writing in their workbooks, and students drawing a diagram. First, the teacher shows a video animation of the Earth and the Sun and highlights that the Earth is 'moving around the Sun' (**Earth's orbit**), that **Earth's tilt** is not changing, and that these together mean the **hemispheres** change their relative position to the Sun (**towards or away**). The teacher sums this up as: 'This is why we end up with opposite seasons. Because of that tilt in our axis puts us in different positions in relationship to the Sun'. As Figure 3.11c shows, this echoes the base cluster of the schematic constellation.

Second, the teacher asks students 'to write one or two sentences that explains how the Sun and the Earth create seasons'. She plays the animation again while students write for several minutes, before soliciting their answers. The first student answer is that 'The tilt of the Earth on a 23.5 degree angle and the orbit around the Sun makes the Earth have seasons', which is to state that **Earth's tilt** and **Earth's orbit** lead to **seasons**. This does not include **hemispheres** or pointing **towards or away** from the Sun and the teacher responds: 'It doesn't quite explain, though, *why* we get the different seasons'. The next two answers provide what she is seeking:

STUDENT   The Earth is always on a 23 ½ degree tilt. This tilt remains the same as we orbit the Sun but the area facing the Sun is different [**towards or away**]. This causes the seasons.

TEACHER   Fantastic. I like that. Very good. Who else has one?

STUDENT   The seasons are created by the Earth's 23.5 degree tilt. When the northern hemisphere is tilted towards the Sun, it is summer. As the Earth orbits the Sun the tilt stays the same but the side [**hemispheres**] that's tilted towards the Sun changes [**towards or away**], making it winter in the northern hemisphere because it's furthest away.

TEACHER   Fantastic, I like that. That's a good one. All right. Hopefully yours says something similar to that.

The first answer adds the node of pointing **towards or away** from the Sun and the second answer adds both that node and **hemispheres**. Once these responses complete the constellation that she had set out, the teacher moves on.

Finally, a simpler version of the constellation is repeated again while the students draw a diagram. The teacher shows a simulation in which she can move the Earth to different positions around the Sun. Students are told to draw a diagram of the Sun and the Earth for one season. 'The key point here', she emphasizes, 'is to make this an accurate diagram'. Through questions, the teacher solicits from students 'the things we cannot be sloppy on in this diagram': 23.5 degrees **tilt** and axis, lines for the equator to show **hemispheres**, and an elliptical **orbit** around the Sun. Thus, the independent nodes of the 'base cluster' are again emphasized as key factors.

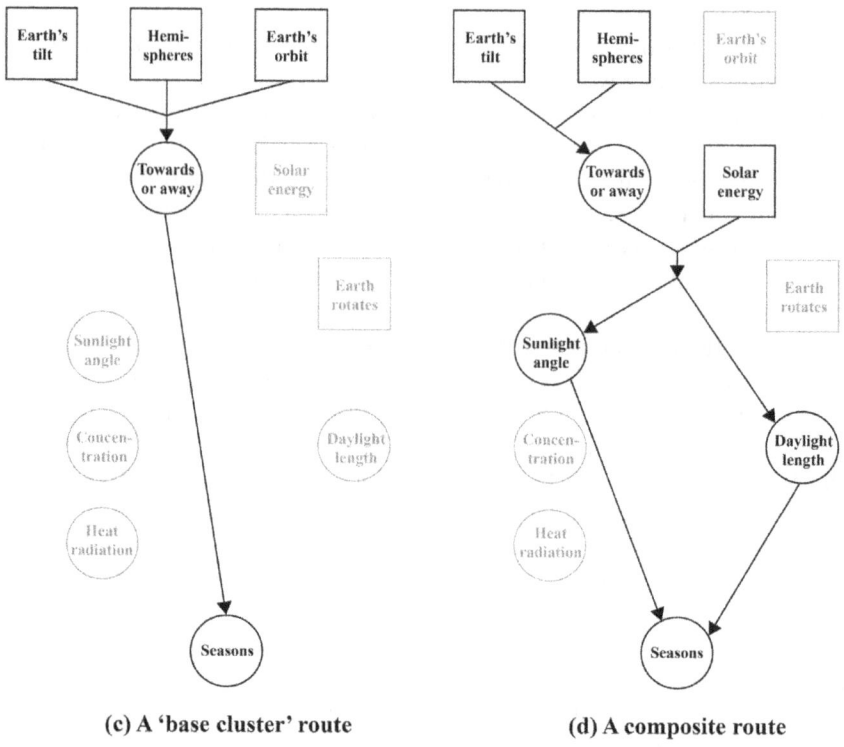

(c) A 'base cluster' route   **(d) A composite route**

FIGURE 3.11 Pedagogic constellations for seasons – 'base cluster' and combined routes

(d) *A composite route*

The final explanation comprises a summary spoken by the teacher:

> We know that because the Earth's tilted, that different parts of the Earth will
> face the Sun at different parts of the year. That difference then in the sunlight
> striking the Earth will lead to different seasonal variations in temperature,
> depending on how much the Earth has heated up during the day. The length
> of our day will also then depend on how far north or south you are from the
> equator. And that explains why temperatures over the year will change at dif-
> ferent locations.

As Figure 3.11d shows this states that **Earth's tilt** means different **hemispheres** –
routinely called 'parts of the Earth' through the lesson – will point **towards or
away** from the Sun, which means that **solar energy** will strike the Earth in
different ways (**sunlight angle**), which leads to variations in temperature (**seasons**), but that this also depends on how much the Earth has heated during the
day due to variations in **daylight length**. Taken together, this all explains the
seasons. So, this explanation includes the 'base cluster' (minus **Earth's orbit**),

her 'sunlight angle' route and her 'daylight length' route. In short, it provides a summarizing composite of the three routes she has taken the class on when creating explanations.

Aggregating explanations

Our concern has been not to evaluate this teaching but to explore what it can tell us about relations among ideas. Here, explaining the seasons seems to offer a variety of potential pedagogic constellations. During an hour-long lesson the teacher takes the class on four different routes, focused on 'daylight length', 'sunlight angle', the 'base cluster' (on which she checks students' understanding), and a composite of all three. This involves what we termed *aggregating*: separate clusters are assembled, each on its own, before being combined. Moreover, this aggregating involves node options: some nodes from the schematic constellation were not included: **sunlight concentration**, **heat radiation** and **Earth rotates**. This is, though, not a gap in teaching. The schematic constellation is not a list of essential ideas but rather a composite of all nodes found in the textbooks we analyzed; not all textbooks include all those nodes. There are likely to be sound pedagogic reasons for employing a certain level of complexity and not adding further nodes that may overcomplicate the explanation.

Conclusion

Science is complex – seemingly simple empirical phenomena may require complex explanations. Our argument has been simple: relations among ideas are one aspect of this complexity and constellation analysis offers a way of seeing these relations and analyzing the roles they play in building knowledge. Our analysis of the logic of explanations of tides and seasons that are offered by textbooks aimed at Year 7 science in NSW, Australia, made the simple point that relations among ideas differ even between otherwise seemingly similar sets of ideas. Our analysis of how those explanations were taught in the same classroom highlighted the simple point that differences in how sets of ideas are taught may be related to differences in relations among their ideas. By analyzing their pedagogic constellations in relation to their schematic constellations, we were not arguing that the logic of explanations is an ideal against which teaching should be measured. As we emphasized, there are many sound reasons for teaching knowledge in different sequences to, or in different degrees of detail than shown by schematic constellations. (Indeed, each textbook unfolds its explanation differently to the composite schematic constellation). Rather, our aim was to show that relations among ideas in an explanation may affect how those explanations are taught – in short, the logic may help shape the pedagogic.

To demonstrate this point, we focused on examining simple distinctions between two kinds of links, clusters, and forms of constellation-building. Using these distinctions we showed that the schematic constellation for tides involves a base cluster that is elaborated successively by supplementary clusters, while that for seasons involves parallel supplementary clusters that 'unpack' part of the base cluster and offer different ways of reaching 'seasons'. Put simply, the elaboration of clusters for tides was akin to 'A then B then C', where for seasons it was akin to 'A and/or B and/or C'.

Using the same distinctions we then analyzed how each explanation was taught by the same teacher during the course of a single unit of study. The pedagogic explanation of tides echoed the logic of its schematic constellation: linear and successive clusters of ideas that build on preceding ideas. The pedagogic explanation of seasons also echoed the logic of its schematic constellation: the teacher set out four different selections and arrangements of ideas that offered differing versions. So the two explanations differed in how they were taught by the same teacher. From this analysis we suggested that 'seasons' appears more amenable to variation of which ideas are selected and how they are brought together. In short, these *assembling* and *aggregating* forms of building constellations differed in ways that echoed relations among ideas within the logic of the explanations.

Beyond our modest aims, these insights suggest that constellation analysis may shed light on relations between research, curriculum and pedagogy. By showing how knowledge changes as it is transformed into curricula or taught in classrooms, constellation analysis could open the 'black box' of recontextualization. Following Bernstein (1990), LCT distinguishes between the logics underlying *production fields* that create 'new' knowledge, *recontextualizing fields* that create curriculum, and *reproduction fields* or sites of teaching and learning (Maton 2014: 43–64). Movements of knowledge between fields are held to involve 'recontextualization' – selection, arrangement and enactment of ideas – that restructures that knowledge. As yet little light has been shed on differences in recontextualization. Constellation analysis offers a way of analyzing how a set of ideas is structured one way in research, differently in a curriculum, and differently again when taught and learned in classroom practice. Comparative analysis could show that some constellations are more amenable to restructuring than others, providing insight into the nature and value of different recontextualizations. Analysis could also reveal when and how it is valuable for the pedagogic unfolding of a constellation to differ from the schematic constellation of its logic.

Constellation analysis may also be practically useful for educators. It offers a way of mapping out lesson plans and teaching designs for the content to be taught and learned. Constellation diagrams could help educators make the knowledge being taught more explicit to students by highlighting key ideas and relations among them. They could be used to map progress through the sequencing of content and make visible how different issues come together. They could also serve as a way for students to demonstrate their understanding. Comparing students' diagrams of, for example, an explanation of tides to the teacher's diagram could help make clear what has been learned and what nodes, links and clusters need revisiting. In this way, constellation analysis offers a means of connecting studies of knowledge and studies of student conceptions, the dominant preoccupation in science education research.

Another area for future development is exploring different kinds of relations among ideas. Links may take many forms. One can draw here on other LCT concepts to examine their attributes. For example, enacting the concept of *epistemological condensation* (specifically its 'translation device' for clausing, see Maton and Doran 2017) would show how different links – classifying, composing, causing, correlating, etc. – add meaning to constellations at different rates. Similarly, analysis of *charging* would show how different valuations of nodes and clusters suffuse constellations. The possibilities offered by such work are yet to be explored. This chapter was but a first step: the future may be written in the stars.

Acknowledgements

We wish to sincerely thank Claire Flanagan for creating Figure 3.1, Greg Rusznyak for creating Figure 3.2 (inspired by http://www.datasketch.es/may/code/nadieh/), and David Fergusson and Kylie Wynne for insights into science textbooks.

Notes

1 See MacDonald (1998) and data available at www.stellarium.org.
2 Rusznyak (2020) goes further by using the constellation of Orion to additionally illustrate the concepts of *semantic density* and *condensation*.
3 Though these concepts were first introduced using concepts from the Semantics dimension of LCT (Maton 2014: 148–70), they do not belong to any dimension. Constellation and cosmological analyses can be conducted with concepts from *any* dimension.
4 Exceptions include: Lambrinos (2020), which distinguishes relations among nodes in dance teaching by how much complexity they add; and Doran (2020), which explores relations in terms of rhetorical moves that show how values are constellated.
5 For a complementary view of relations among ideas from systemic functional linguistics, see Doran and Martin, Chapter 5 of this volume.
6 Aggregating may involve separate clusters or separate nodes. In this chapter, teaching about Earth's seasons involves several clusters of ideas that are built separately before being brought together. In other fields aggregating may involve a large number of individual nodes. For example, in History lessons the discussion of a period may involve aggregating a large number of facts (Maton 2015).
7 Both are also *epistemological constellations*, in which stances are formal definitions (such as concepts and empirical descriptions). For discussion of *axiological constellations*, in which stances are affective, aesthetic, ethical, moral or political, see Maton (2014: 148–70), Martin *et al.* (2010), Doran (2020), and Tilakaratna and Szenes (2020).
8 We used the same method on the same textbooks for both tides and seasons. We expand on the process for constructing 'schematic constellations' in a future publication.
9 We draw on a major study funded by the Australian Research Council (DP130100481).
10 The video (https://www.youtube.com/watch?v=QcbN9SVkqYU) is by the US public service broadcaster KQED; the excerpt is narrated by Ben Burress, staff astronomer at the Chabot Space and Science Center in Oakland, California.
11 As stated earlier, our pedagogic constellation diagrams show nodes and links *in comparison to the schematic constellations* because of our specific aims here. Constellation diagrams of the teaching as it unfolds would differ.

12 The two preceding lessons included mentions of Earth's tilt, orbit, rotation, and how sunlight heats the Earth. However, these ideas were nodes in other constellations: explaining 'day and night' and explaining different temperatures on Earth. In neither case was the content constellated into an explanation of *seasons*. This highlights that no content is locked into a specific constellation. Mentions of, say, 'tilt' or 'orbit' are not necessarily discussions of 'seasons'.

13 The video is available at: https://www.teachertube.com/videos/what-causes-seasons-on-earth-657. Maton and Howard (Chapter 4, this volume) examine the teacher's use of this video in detail, as an example of using multimedia in science teaching, explaining why she replaced its audio with her own commentary.

References

Adendorff, H. and Blackie, M. A. L. (2020) 'Decolonizing the science science: When good intentions are not enough', in C. Winberg, S. McKenna and K. Wilmot (eds) *Building Knowledge in Higher Education*, London: Routledge, 237–54.

Arena, P., Warnant, P., Burrows, K., Everygreen, M. J. and Lofts, G. (2009) *Core Science: Stage 4 Complete Course*, Milton: John Wiley and Sons.

Bernstein, B. (1990) *Class, Codes and Control, Volume IV: The Structuring of Pedagogic Discourse*, London: Routledge.

Bishop, S., Bass, G., Champion, N., Gregory, E., McKenna, E. and Walker, K. (2011) *Nelson iScience for the Australian Curriculum, Year 7*, South Melbourne: Nelson Cengage Learning.

diSessa, A. (1993) 'Toward an epistemology of physics', *Cognition and Instruction*, 10(2/3): 105–225.

Doran, Y. J. (2020) 'Cultivating values: Knower-building in the humanities', *Estudios de Lingüística Aplicada*, 37(70): 169–98.

Ellery, K. (2017) 'Conceptualising knowledge for access in the sciences', *Higher Education*, 74(5): 915–31

Georgiou, H., Maton, K. and Sharma, M. (2014) 'Recovering knowledge for science education research: Exploring the "Icarus effect" in student work', *Canadian Journal of Science, Mathematics, and Technology Education*, 14(3): 252–68.

Glenn, E. (2016) '*From clashing to matching: Examining the legitimation codes that underpin shifting views about climate change*', unpublished PhD thesis, University of Technology Sydney, Australia.

Lambrinos, E. (2020) '*Building ballerinas: Developing dance and dancers in ballet*', unpublished PhD thesis, University of Sydney, Australia.

Land, R., Cousin, G. and Meyer, J. H. F. (2005) 'Threshold concepts and troublesome knowledge (3): Implications for course design and evaluation', in C. Rust (Ed.) *Improving Student Learning Diversity and Inclusivity*, Oxford: OCSLD, 53–64.

MacDonald, J. (1998) *The Arctic Sky: Inuit Astronomy, Stare Lore, and Legend*, Ontario: Royal Ontario Museum.

Martin, J. R., Maton, K. and Matruglio, E. (2010) 'Historical cosmologies: Epistemology and axiology in Australian secondary school history discourse', *Revista Signos*, 43(74): 4330–463.

Maton, K. (2014) *Knowledge and Knowers: Towards a Realist Sociology of Education*, London: Routledge.

Maton, K. (2015) '*Legitimation Code Theory: Past, present, future*', Opening Address, *LCT1 – First International Legitimation Code Theory Conference*, Cape Town, South Africa, June.

Maton, K. and Doran, Y. J. (2017) 'Condensation: A translation device for revealing complexity of knowledge practices in discourse, part 2 – clausing and sequencing', *Onomázein Special Issue*, March: 77–110.

Meyer, J. and Land, R. (2003) 'Threshold concepts and troublesome knowledge', *ETL Project Occasional Report 4*, http://www.etl.tla.ed.ac.uk/docs/ETLreport4.pdf

Rickard, G. (2011) *Pearson Science 7: Student Book*, Frenchs Forest: Pearson Education.

Rose, D., McInnes, D. and Korner, H. (1992) *Scientific Literacy*, Sydney: Metropolitan East Disadvantaged Schools Program.

Rusznyak, L. (2020) 'Supporting the academic success of students through making knowledge-building visible', in C. Winberg, S. McKenna and K. Wilmot (eds) *Building Knowledge in Higher Education*, London: Routledge, 90–104.

Salwén, H. (2019) 'Threshold concepts, obstacles or scientific dead ends?', *Teaching in Higher Education*, DOI: 10.1080/13562517.2019.1632828

Stannard, P. and Williamson, K. (1995) *Science World 7*, South Yarra: Macmillan.

Tilakaratna, N. and Szenes, E. (2020) '(Un)critical reflection: Uncovering disciplinary values in social work and business reflective writing assignments', in C. Winberg, S. McKenna and K. Wilmot (eds) *Building Knowledge in Higher Education*, London: Routledge, 105–25

Unsworth, L. (1997) '"Sound" explanations in school science: A functional linguistic perspective on effective apprenticing texts', *Linguistics and Education*, 9(2): 199–226.

Zhang, J., Alford, D., McGowan, D. and Tilley, C. (2013) *Oxford Insight Science: Australian Curriculum for NSW, Stage 4, Year 7*, South Melbourne: Oxford University Press.

4

ANIMATING SCIENCE

Activating the affordances of multimedia in teaching

Karl Maton and Sarah K. Howard

Integration is more than selection; affordances are more than interactions.

Introduction

Multimedia objects that combine diverse visual and audio elements – text, pictures, moving images, speech, sounds, etc. – are a common feature in classrooms (Li *et al.* 2019). Increased access to the Internet and rapid growth in online 'educational' materials has made such multimedia as animations evermore available to teachers as classroom resources (Berney and Bétrancourt 2016). These are particularly of interest to science education as ways of displaying complex explanations (Ploetzner and Lowe 2012). Moreover, studies claim a 'multimedia effect' whereby students learn better through words and pictures together (Mayer 2003). However, not all multimedia are created equal – some are more suited to classroom practice than others. Moreover, even if designed for a particular curriculum, a multimedia object is unlikely to match the needs of a specific task in a specific lesson in a specific classroom. Thus, questions of how teachers integrate such ready-made objects into classroom practice are increasingly significant (e.g. Jenkinson 2018). Yet, these questions are not the concern of most education research into multimedia. Instead, the principal concern is exploring *cognitive* processes of *student learning* and generating principles for *designing* multimedia that support those processes (e.g. Mayer 2014b, Mutlu-Bayraktar *et al.* 2019). Questions of how *pedagogic* practices by *teachers* support *integrating* multimedia into classroom tasks are marginal. This chapter contributes to foregrounding and addressing these questions in the context of science teaching.

We begin by arguing that to understand integration requires bringing into the picture both teaching and the knowledge practices being taught. To begin meeting this need we draw on the notion of 'affordances' (e.g. Bower 2017), which highlights that technology such as multimedia objects differ in their abilities to meet the demands of different classroom tasks. However, we argue that current uses of 'affordances' are too limited. Studies overwhelmingly examine affordances for inter- actions, especially among students, and ignore affordances for teaching and learn- ing specific knowledge practices. To address this 'knowledge-blindness' (Howard and Maton 2011, 2013), we introduce concepts from the Autonomy dimension of Legitimation Code Theory. These concepts reveal one aspect of the knowledge practices expressed by resources and their *epistemic affordances* for teaching and learn- ing. Specifically, *positional autonomy* conceptualizes relations between the content required by a task and that expressed by a resource, *relational autonomy* conceptual- izes relations between the purposes for which that content is needed by a task and those for which it is used in a resource, and *target* places the specificities of each classroom task at the centre of analyzing these two relations. Enacting these con- cepts, we explore both the affordances of multimedia objects (and other resources) for building knowledge in tasks and how teachers select and activate those epistemic affordances in classroom practice.

Specifically, we conduct two in-depth analyses of integrating multimedia in sec- ondary school science classrooms. We focus on the use of animations, a central concern of multimedia research (Li *et al.* 2019). In both examples, a science teacher selects an animation whose affordances are well suited to their task, both in terms of interactions enabled and knowledge practices expressed. However, while one teacher activates its epistemic affordances, the other fails to do so. These analy- ses illustrate that integration does not end with selection and that *pedagogic work* by teachers is required to activate the epistemic affordances of resources. In short, they show how understanding success in teaching science with multimedia requires seeing not only the multimedia but also the teaching practice and the science knowledge being taught.

Integrating multimedia into teaching

The cognitive–learning–design axis

Multimedia combine words (spoken or text) and pictures (such as still or moving images). It is widely acknowledged that multimedia resources are becoming more accessible for education (Berney and Bétrancourt 2016). How, then, can teachers best integrate such multimedia objects as animations into classroom practice to sup- port teaching science? This is our focus in this chapter. However, this is not the focus of the vast majority of research into multimedia (see Li *et al.* 2019). Most studies focus on learning rather than teaching, designing rather than integrating multime- dia, and computer-based contexts rather than classrooms (e.g. Mayer 2014b). This is not to dismiss this body of work but rather to highlight that dominant approaches to the educational potential of multimedia are not concerned with integration.

Instead the focus of most research lies on a cognitive–learning–design axis. Commonly used models, such as 'cognitive theory of multimedia learning' (Mayer 2014a) and 'cognitive load theory' (Sweller *et al.* 2019), examine such issues as how the brain processes information, audio and visual channels, and kinds of memory. Implications are drawn for how to combine visual and auditory information in order to, for example, manage 'cognitive load' (e.g. Mayer and Moreno, 2003). This provides insights into design principles, such as the 'spatial congruity principle' that text and visual content should be proximate (Mayer and Fiorella 2014). However, this cognitive–learning–design focus obscures integration in two principal ways. First, teachers and teaching are treated as if limited to designing 'environments' or selecting resources that meet design principles. This leaves aside almost all real-world classroom pedagogy. Second, the forms taken by the knowledge practices to be taught and learned are absent or reduced to the status of 'topic', 'subject matter' or 'information'. This ignores that learning involves learning something and that the forms taken by the knowledge practices expressed by multimedia may affect their integration into teaching and learning of that something. Thus, while valuable for developing generic principles for designing multimedia, these frameworks are not well suited to questions of integrating multimedia into teaching specific knowledge. We thus need to bring teaching and knowledge into the picture.

The affordances of 'affordances'

A fruitful starting point is the notion of 'affordances' which is typically used in education research to explore the capacities of technologies for enabling actions (Bower 2008, Hammond 2010, Antonenko *et al.* 2017). For classroom practices, this is to say that a resource may be more suited to some tasks than others. Affordance 'frameworks' list 'abilities' of technology; Bower (2017), for example, includes read-ability, write-ability, playback-ability, share-ability, among others. The aim is for 'designers' of 'learning environments' to select resources that offer the abilities required by tasks. The notion of 'affordance' thus offers the potential to foreground classroom tasks and so integration.

However, this affordance of 'affordance' is often not activated by research, which is animated by the cognitive–learning–design axis. For example, the notion is said to help show 'how tools interplay with *cognition* and hence how to best *design* educational systems that meet the *learning* requirements of tasks' (Bower 2008: 4; emphases added). Studies also share the blind spots of teaching and knowledge practices. Consider the proclaimed affordances of 'affordance'. Hammond (2010: 211), for example, states: 'This is its major value as a concept: it is not the tool, it is not the person, it is the interaction of tool and person'. Focusing on interactions that a 'tool' enables, both between 'tool and person' and among people, is indeed the principle way that 'affordances' is used. However, the people studied are only learners and not teachers, the 'interaction' is only learning and not teaching, and the role played by what is to be taught and learned through the interaction is absent. This focus on student interactions is underscored by the huge field of 'Computer-Supported Collaborative Learning', whose gravitational weight pulls attention away from real-world classrooms and most forms of pedagogy. For example, affordances are often clustered together as supporting, for example, 'static/instructive' or 'collaborative/

productive' interactions (Bower 2017), where the latter focus on student interactions garners most research.

In short, existing research focuses on what we shall term 'interactive affordances'.[1] This underestimates the affordances of the notion as an animating metaphor for research. To help examine integration, we shall focus on *epistemic affordances* – how knowledge practices expressed by a resource offer opportunities for teaching and learning – and analyze how teachers can activate those epistemic affordances in teaching science.[2] To do so, we draw on Legitimation Code Theory.

Legitimation Code Theory: Autonomy

Legitimation Code Theory (LCT) is a framework for researching and shaping all kinds of practice (Maton 2014) but has proven particularly valuable for exploring knowledge practices (e.g. Howard and Maton 2011). LCT is also widely enacted to analyze and inform teaching.[3] The framework comprises several *dimensions* that reveal different aspects to knowledge practices. For example, concepts from the Semantics dimension (Maton 2020) can, among other things, reveal how the context-dependence of meanings expressed by a resource can support a task, such as affording the desired degree of concreteness.[4] Concepts from the Specialization dimension (Maton 2014) can reveal, among other things, how the representational fidelity of images afford possibilities for teaching and learning specific ideas. Indeed, LCT affords far more possibilities for analysis than can be enumerated here. Thus, we shall focus on the broader issue of the content and purpose of knowledge practices expressed by resources. We draw on the Autonomy dimension, which is particularly suited to exploring integration (Maton and Howard 2018, chapter 2 of this volume).

Autonomy

The dimension of Autonomy explores what makes practices distinctive. It begins from the simple premise that any set of practices comprises constituents that are related together in particular ways. Autonomy explores how practices establish different degrees of insulation around their constituents and around the ways they are related together. Here we can describe constituents as the *content* of practices and how they are related together as the *purpose* to which they are put. We can then distinguish:

- *positional autonomy* (PA) between content within a context or category and content from other contexts or categories; and
- *relational autonomy* (RA) between the purposes to which content is put within a context or category and purposes from other contexts or categories.

Each may be stronger or weaker along a continuum of strengths. The stronger the positional autonomy, the more insulated are contents from those of other contexts or categories; and the stronger the relational autonomy, the more insulated are purposes from those of other contexts or categories.

TABLE 4.1 Generic translation device, first two levels (adapted from Maton and Howard 2018: 10)

PA/RA	1st level	PA/RA	2nd level
stronger	target	++	core
		+	ancillary
weaker	non-target	−	associated
		− −	unassociated

The 'contexts or categories' in these definitions are held open to embrace the manifold diversity of practices. To enact the concepts, one needs to develop *translation devices* that show how they are realized within specific problem-situations (Maton and Chen 2016). As shown in Table 4.1, Autonomy offers a *generic translation device* as a first step. This device is activated by asking: what contents and purposes are considered constitutive of *this* context or category, here, in *this* space and time, for *these* actors? This gives a 'target' against which to compare practices. If their contents are from within this target, they embody stronger positional autonomy, and if from elsewhere, they embody weaker positional autonomy; and the same holds for purposes and strengths of relational autonomy. A more fine-grained analysis may then ask: what of the target is considered *core* and what *ancillary,* and what of other contents and purposes are considered *associated* with or *unassociated* from the target? As Table 4.1 shows, this generates four strengths for positional autonomy and for relational autonomy.

This device is, though, still 'generic'. To translate to specific data we must decide whose target to examine and at what level. First, a target is always *someone's* conception of what makes a context or category distinctive; whose target is chosen to ground analysis depends on the problem-situation. We shall analyze two examples of teaching, so we focus on each teacher's targets (and use such possessives throughout the chapter).[5] Second, targets can describe 'contexts or categories' of all kinds. For example, in our analysis of integrating mathematics into science (chapter 2, this volume), each teacher's target is a syllabus stage and their core target is a unit of study. These broad categories capture relations between two subject areas across whole lessons. Here our questions are more fine-grained: how a specific multimedia object relates to specific tasks in specific lessons. Thus, we analyze each teacher's target as the lesson and their core target as the task in that lesson. Using the translation device of Table 4.2, we can revisit the concepts to state that here:

- strength of *positional autonomy* conceptualizes whether contents expressed by a multimedia object match the teacher's core target content for a task (PA++), match their target content for the wider lesson (PA+), come from beyond the lesson (PA−) or come from beyond the unit (PA− −); and

TABLE 4.2 Translation device for analysis in this chapter

PA/RA	1st level	In this analysis:	2nd level	In this analysis:
++	*target*	lesson	*core*	task
+			*ancillary*	rest of lesson
−	*non-target*	other contents or purposes	*associated*	other unit lessons or related to task
− −			*unassociated*	knowledge from beyond unit

- strength of *relational autonomy* conceptualizes whether the purposes of a multimedia object match the teacher's core target purposes for the task (RA++), match their target purposes for the lesson (RA+), come from beyond the lesson (RA−) or come from beyond the unit (RA− −).

More succinctly, *positional autonomy* will show relations between the content expressed by a multimedia object and that required by a teacher's task, and *relational autonomy* will show relations between the purposes for which that content is used in a multimedia object and the purposes needed by a teacher's task.

We shall illustrate the usefulness of these concepts through analyses of two teachers using video animations in science classrooms.[6] Both examples are from Year 7 of secondary schooling in New South Wales, Australia. Both are from a unit of study called 'Earth and space sciences'. We develop specific translation devices for each analysis, showing contents and purposes for each task. Given that multimedia objects are complex sets of diverse elements, we analyze their key elements separately. We map these elements on an *autonomy plane*, as shown in Figure 4.1, divided into the 16 modalities generated by combining the strengths of positional autonomy and relational autonomy outlined above. By locating elements of each animation in relation to the teacher's targets, we show its epistemic affordances for the task, the pedagogic work required by the teacher to integrate the animation into the task, and whether the teacher succeeds.[7] By 'pedagogic work' we mean here changing the positions of elements on the plane. This can involve: *integration* or connecting to content (strengthening PA, or moving up on Figure 4.1); *disintegration* or disconnecting from content (weakening PA or moving down); *introjection* or turning to purpose (strengthening RA or moving right); and *projection* or turning to another purpose (weakening RA or moving left). We can also describe *addition*, where new contents or purposes are added, *subtraction*, where they are removed, and *substitution*. Through such changes, teachers and students can more closely match the knowledge practices of multimedia and tasks.

As we shall show, both teachers select animations whose interactive affordances match the interactional needs of their tasks and whose epistemic affordances largely

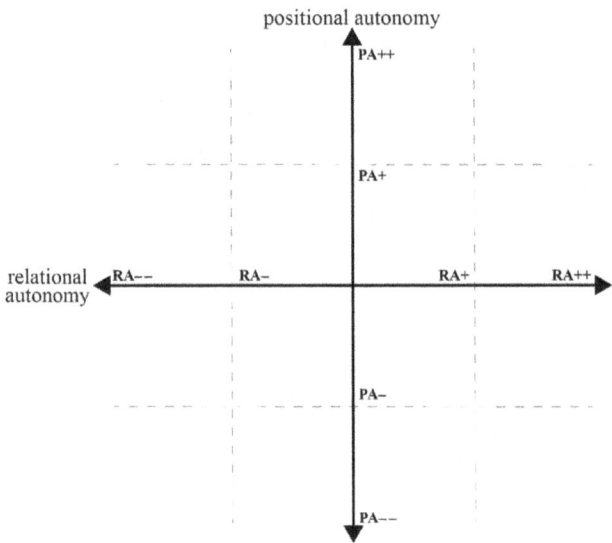

FIGURE 4.1 The autonomy plane (divided into 16 modalities)

match the knowledge practices they intend to teach. They are well suited but, as is typical for classroom practice, not perfect. In our first example, the animation's epistemic affordances need pedagogic work to activate their potential to support the task but the teacher does not provide that integrative pedagogy. In the second example, the teacher's pedagogic work makes the animation a closer match to the knowledge needs of her task by substituting the animation's audio elements with her own commentary. Both animations have limited interactive affordances but plentiful epistemic affordances, highlighting that these are not confined to interactions and require activating. The analyses thereby bring knowledge and teaching to the fore when examining integration.

Failing to activate epistemic affordances

The teacher's targets: Shooting for the Moon

In our example of teaching that fails to activate epistemic affordances, the teacher declares his target for the lesson at its outset: 'today we are going to continue on with the Moon'. For the task in question, the content of his core target concerns the Moon's rotation and its purpose is for students to answer two questions:

> I'm going to pose a question for you … [reading PowerPoint slide] 'Rotation of the Moon'. So write that down: 'Rotation of the Moon' … That's our topic. And then I want you to answer: does the Moon rotate on its axis?. If so, how fast does it rotate?

TABLE 4.3 Specific translation device for Moon animation task

target	The Moon in this lessson	++	*core*	Explanation in task, involving: Moon's orbit, length of orbit, rotation, length of rotation, synchronous rotation, facing side and 'dark side'
		+	*ancillary*	Other science knowledge concerning the Moon expressed in lesson (e.g. phases and eclipses)
non-target	Other knowledge	–	*associated*	Other science knowledge related to the Moon not included in lesson (e.g. tides) or topics in other lessons of the unit (e.g. seasons)
		– –	*unassociated*	Knowledge from beyond the unit

As summarized in Table 4.3, the teacher's *core target* concerns 'synchronous rotation'; in other words, that the time the Moon takes to rotate on its axis is almost the same as the time it takes to orbit the Earth, so we always see the same side. The rest of the lesson focuses on phases of the Moon and eclipses – these form the teacher's *ancillary target*. Other science knowledge related to his target, such as the Moon's role in tides, and from other topics in the same unit, such as seasons, form his *associated non-target*. Knowledge from beyond the unit is *unassociated non-target*.

The Moon animation

The task has two parts: discussion of a collage of static photographs and a video animation of 2:15 minutes length called 'Synchronous rotation'.[8] The animation comprises captions and two-dimensional images of entities, of which one (the Moon) moves. There is no sound. The animation has two main parts: simulations of different rotations of the Moon while orbiting the Earth; and a simulation of the changing shadow on the Moon created by sunlight during its orbit.

In the first main part a half-red and half-white Moon travels around an ellipse four times under the same top caption: 'Do we always see one side of the Moon?'. As shown by Figure 4.2, the first orbit is sub-captioned 'Moon without any rotation…' and the division between its red and white halves remains vertical during its orbit. After the orbit, the sub-caption is extended by: 'we see all sides in this case'. The second orbit is sub-captioned 'Moon with rotation' and the Moon rotates very quickly. As shown in Figure 4.3, after this orbit the words 'How fast is it rotating?' appear and a sub-caption states: 'This **isn't** actually what the Moon does…'. The third orbit is also sub-captioned 'Moon with rotation…' and the Moon spins so that its red side is always facing Earth. Central captions then appear: 'How fast is it rotating? It rotates on its axis **once** in the **SAME** time it takes to orbit us once!'.

FIGURE 4.2 Moon animation: First orbit, without rotation (at 14 seconds)

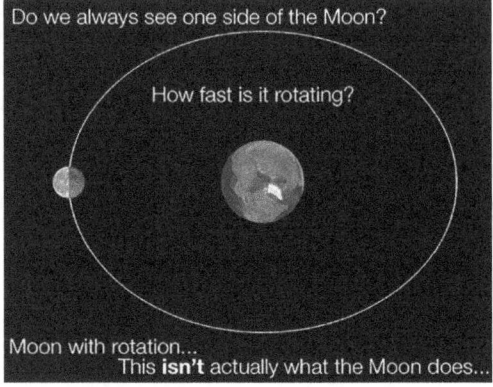

FIGURE 4.3 Moon animation: Second orbit, overly fast rotation (at 53 seconds)

To 'Moon with rotation…' is added 'we only see the red coloured side!!'. The Moon then repeats that orbit.

In the second main part, illustrated by Figure 4.4, the Sun is represented on the left. The top caption changes to 'Now again with the Sun's shadow' and a central caption states 'THE ROTATION PERIOD OF THE MOON IS EXACTLY THE SAME AS ITS PERIOD OF REVOLUTION!'. As it orbits, the right-hand side of the Moon remains darker. Another central caption appears: 'SYNCHRONOUS ROTATION! WILL EVENTUALLY HAPPEN TO THE EARTH TOO! (IT'S [sic] ROTATION RATE WILL SLOW DOWN TO EVENTUALLY MATCH ITS REVOLUTION RATE AROUND THE SUN)'. Finally, a sub-caption states: 'We always see the red coloured side! Position of the Sun's shadow does not move'.

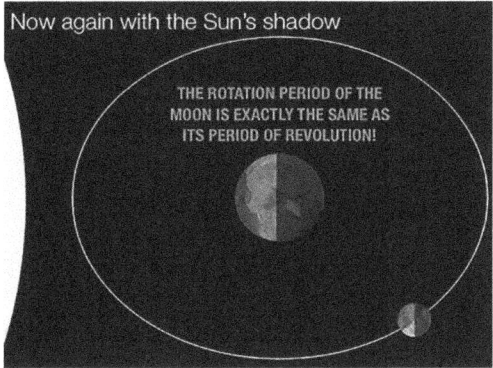

FIGURE 4.4   Moon animation: Sun's shadow simulation (at 1:53 minutes)

Affordances for the task

In terms of interactive affordances, the animation is relatively limited. It is freely accessible online, shareable, watchable, viewable (though not listenable), and can be played back. The animation is not easily manipulated, beyond pausing, repeating or jumping ahead. It does not lend itself to being redrawn or support actions such as drawing, writing or recording. These interactive affordances are associated by Bower (2017) with 'static/instructive' rather than 'collaborative/productive' interactions. Though limited in interactivity, the animation can support the teacher's stated purpose: students answering two questions about the topic.[9]

Epistemic affordances

To examine its epistemic affordances, Figure 4.5 presents the animation's elements on the autonomy plane, using the translation device of Table 4.3. (Positions *within* each modality do not denote differences in strengths). To give an indication of prominence, the shading offers a heuristic heat map of relative duration overall, from almost ever-present (darkest) through prolonged and briefer to momentary (lightest).[10]

As Figure 4.5 summarizes, the animation visually represents the teacher's core target knowledge: key factors – such as lengths of orbit and rotation – are animated at length to explain synchronous rotation (PA++, RA++). However, the animation also spends prolonged periods outside the teacher's target. The first minute comprises orbits with no rotation and with overly fast rotation, both of which are inaccurate. In terms of content, they are not part of the explanation (PA–). Whether they serve the purpose of learning about synchronous rotation depends on how this inaccuracy is presented. For the first orbit with no rotation, its inaccuracy is not made explicit. The caption asks whether we always see one side of the Moon, rather than stating that as a fact, and there is no indication that the viewer should judge whether, given that fact, the animated rotation is correct or not. Indeed, the caption implies the opposite: that the viewer should use the animated rotation to answer the question of whether we only see one side of the Moon. This orbit is thus not

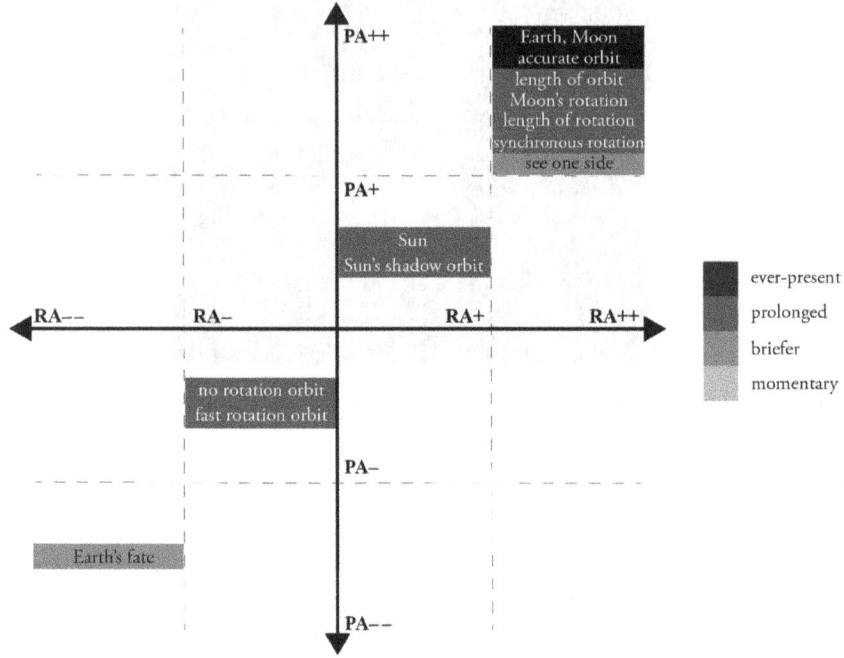

FIGURE 4.5 Autonomy analysis (with heuristic heat map) of Moon animation

serving the purpose of teaching synchronous rotation; it exhibits both non-target content and non-target purpose (PA–, RA–). The second orbit, with overly fast rotation, exhibits the same non-target content and purpose (PA–, RA–): only after the orbit ends and a caption asks how fast it is rotating does the animation make its inaccuracy explicit.

The third orbit represents the Moon's rotation accurately. Captions then ask and answer the teacher's second question of 'how fast' the Moon rotates: core target knowledge (PA++, RA++). During the final simulation showing the shadow on the Moon created by the Sun, this knowledge is further underlined by captions. The changing shadow itself is part of the teacher's wider target for the lesson: phases of the Moon (which the shadow demonstrates) is the next topic he discusses – ancillary target knowledge (PA+, RA+). Further from his target lies the caption about the Earth's fate, which is not part of the unit (PA– –, RA– –).

In summary, the animation's interactive affordances can support the teacher's 'instructive' task of having students consider two questions. The animation also exhibits significant epistemic affordances for the task. On both counts, then, it appears well selected. However, integration is more than selection. Our autonomy analysis shows that activation of its epistemic affordances requires pedagogic work by the teacher, specifically through *additions*. First, he needs to establish a shared understanding that the same side of the Moon is always seen from Earth and to clarify that the question is whether, given that fact, the rotation being simulated is accurate or not.[11] These additions would support the *introjection* or turning to purpose of the first two (inaccurate) orbits, strengthening their relational autonomy (and so moving them right on the plane to become PA–, RA++). Second, he could explicitly relate the simulation of the Sun's shadow (PA+, RA+) to his next task of discussing phases of the Moon. This addition would support both the *introjection* (turning to purpose, moving right) and the *integration* (connecting to content, moving up) of these elements into his core target (PA++, RA++). Doing so would expand his task to include setting up his subsequent discussion of Moon phases, making the task less discrete and the lesson overall more cumulative.

The dark side of the Moon

The teacher does not undertake the pedagogic work needed to activate the animation's epistemic affordances. He begins by asking students to answer his two questions by looking at a collage of pictures of the Moon from Earth:

> So, there are some pictures taken from all around the world, different days, alright? See if you can work out whether or not the Moon spins on its axis or not, and if it does, how fast it spins. So talk to the person next to you.

A student asks whether 'fast' means 'How many kilometres an hour?', but is not answered. Another student asks 'How are we supposed to know how fast it rotates if we've only got pictures?', to which the teacher's responds 'Well, if you think it turns once every night....'. Several minutes later, he tells the whole class:

> I'll give you a clue: the pictures, if you look very closely, and I've chosen these because they all show it, you just have to look very closely, it will answer the first part of the question for you. But you have to be careful and I'll explain why.

When he asks students to report back, several respond by stating that the Moon does spin. The teacher claims that a 'dot' appears on the Moon in every picture (though cannot point to it on several) and says this 'dot' shows that it spins. Thus what could establish that we always see the same side of the Moon – that pictures 'taken from all around the world, different days' have the same 'dot' – is instead used to state that the Moon rotates. On 'how fast' it spins, a student answers: 'Is it the same speed as the spin on its orbit? ... It turns the same as the speed ... like the speed of when the Moon turns, it's the same as the speed... the Moon rotating'. The teacher responds:

> Yeah, alright. Did everybody understand? ... Now, let's watch a video to explain why it spins on its axis once every twenty-nine and a half days.[12]

So, the teacher has not added to the animation that: the same side of the Moon is always seen from Earth; this fact should be used by students to judge whether the animation's simulations are accurate; and several simulations will be inaccurate. Instead he plays the first (inaccurate) orbit and says: 'Now, it's just a visual with some … but you actually have to watch it, okay? There's no sound, just watching'. After the first orbit he pauses the video and says 'Alright, so that's what happens. We'll play that again. Alright? We'll play that again'. He thus presents the inaccurate orbit as accurate.

The teacher returns to the start and states that the red 'side' of the Moon is 'the dark side' and 'the side that you actually see'. He asks 'what happens if it doesn't spin?' and plays the first orbit again. These additions still do not establish that only one side is visible from Earth and that simulations may be inaccurate. The students remain unusually silent and the teacher answers his own question: 'we can see the opposite side of the Moon… but we don't ever see the opposite side of the Moon, so that can't work'. To this a student loudly exclaims 'What?!'. The teacher then plays the second orbit (fast rotation) while saying 'Now, so again you'd see all sides of the Moon, so that's not the case'. A student responds hesitantly: 'I don't think… does the Moon spin like that?'. As the third (correct) orbit plays, a student adds 'It seems the red side is always facing us', to which the teacher confirms 'The red side *is* always facing the Earth … So yeah, it takes – every time it orbits the Earth, it spins once on its axis'. Whether by 'red side' he means in the animation or reality is not clear. After the teacher says this is what the student was saying in his answer prior to the animation, another student loudly exclaims: 'There's only one person understanding! There's only one person understanding!'.

The teacher pauses the animation and tells students to write down a caption as their answer: 'It rotates on its axis **once** in the **SAME** time it takes to orbit us once!'. After two minutes he continues the animation, reading out a caption that calls the phenomenon 'synchronous rotation'. The teacher ignores the Sun's shadow simulation and eschews the opportunity to connect the animation with phases of the Moon. He then opens a new PowerPoint slide and begins discussing that topic separately.

In summary, the animation offers much that is needed for the teacher's task. He activates some of the animation's limited affordances for interactivity, principally pausing and replaying. However, while doing so he deactivates the animation's epistemic affordances. Instead of adding knowledge that could turn to purpose the simulations of its first part, introjecting 'no rotation orbit' and 'fast rotation orbit' rightwards on Figure 4.5 (to become PA−, RA++), he suggests the first orbit is correct and within his core target, replaying the orbit as if it was accurate. What is and is not core target content is thus unclear. As illustrated by the outbursts of students expressing confusion, it is likely many remained in the dark about the Moon's synchronous rotation. Second, he does not add to the second part (on the Sun's shadow) anything to activate its affordance for connecting to his next topic. Thus the task remains segmented from other knowledge he discusses in the same lesson.

Activating epistemic affordances through pedagogic work

The teacher's targets: The seasons

To illustrate how a multimedia animation can be successfully integrated into teaching a scientific explanation, we turn to a different teacher at a different school but
teaching the same unit ('Earth and space sciences') in the same curriculum (New
South Wales) at the same level (Year 7 secondary school). The teacher makes her
target for the lesson explicit at the outset:

> The last time I saw you, we were talking about day and night, we were talking
> about the tilt of the Earth on its axis, and we had started to discuss seasons.
> That's what we're going to be looking at today.

As she tells students, the core target content for the task is how 'the tilt of our axes
and our position around the Sun' shape Earth's seasons and her core target purpose
is to support their understanding of that explanation:

> If you'd like to make some notes of things that you might think are important,
> it's a good time. Later on, we're going to ask some questions to see how much
> you understand.

Scientific explanations can become extremely complex. Not all factors in an explanation may be taught in a particular syllabus, year of schooling, unit or task. 'The
seasons' is one such explanation – not everything on the topic necessarily lies within
this teacher's targets. A 'constellation analysis' of this lesson by Maton and Doran
(chapter 3, this volume) shows the factors involved in this specific task to be: Earth's
axis is tilted, Earth has hemispheres, Earth orbits the Sun, that bringing those three
together means the hemispheres point towards or away from the Sun through the
year, the Earth receives energy from sunlight, that pointing towards or away the
Sun means the angle of that sunlight varies, and that this creates the variations in
temperature through the year known as 'seasons'. As summarized in Table 4.4, this
summarizes the teacher's *core target,* both its content (those factors and relations) and
purpose (teaching those factors and relations) – it shows what she means by 'the tilt
of our axes and our position around the Sun'.

Maton and Doran (Chapter 3) also show that the teacher discusses other factors
in different explanations of the seasons she gives during the rest of the lesson, specifically: that the Earth rotates, that pointing towards or away from the Sun while the
Earth rotates leads to variations in the length of daylight, and that changes in daylight
length helps explain the seasons. These factors form her *ancillary target* (see Table 4.4).
Her *associated non-target* comprises other topics in the unit, such as the Moon, or further factors explaining the seasons not discussed in the lesson, such as how variations
in sunlight angle create variations in the concentration of sunlight in each hemisphere.[13] Finally, *unassociated non-target* knowledge is anything else outside her target.

TABLE 4.4 Specific translation device for seasons animation task

target	Explanations of the seasons in this lesson	++	*core*	Explanation in task, involving: Earth's tilt, hemispheres, orbit, hemispheres point towards or away from Sun, solar energy, variations in angle of sunlight, variations in temperature called 'seasons'
		+	*ancillary*	Factors for seasons in other explanations in lesson (e.g. Earth rotation and length of daylight).
non-target	Other knowledge	–	*associated*	Other science knowledge related to seasons not covered in lesson (e.g. sunlight concentration) or topics in other lessons of the unit (e.g. the Moon)
		– –	*unassociated*	Knowledge from beyond the unit (e.g. door shutting sound effect)

The seasons animation

The multimedia object is a two-minute long animation called 'What causes Earth's seasons?' that includes entities, movements, visual effects, captions, labels, spoken narration (a female human voice with American accent), music and sound effects.[14] The animation can be distinguished into four main explanatory stages: (i) focusing the question of what causes Earth's seasons on the role of axial tilt; (ii) outlining an explanation; (iii) showing how this leads to seasons changing through the year; and (iv) summing up.

The animation begins by asking as voice and title text: 'What causes Earth's seasons?'. Electronic, 'jazzy' music continues throughout the animation. Patterned panels leave the title screen to the sound of a heavy door shutting. A photorealistic Earth rotates (to whirring sounds) and becomes abstracted, as illustrated by Figure 4.6, to display continents, latitude and longitude lines, and an axis line. At the same time the voice answers itself: 'Earth's seasons are caused by Earth's tilt on its axis. Instead of going straight up and down [a green vertical line appears briefly through the Earth], Earth's axis tilts 23.5 degrees'. Then the Earth orbits the Sun (while orbited by the Moon), as the voice asks: 'Can you see why the tilt causes Earth to have seasons throughout the year? Let's find out!'.

Second, over a static, rotating Earth, as illustrated in Figure 4.7, the narration outlines how tilt affects the angle of sunlight:

> Here's the energy of the Sun [rumbling sound as yellow band extends from Sun to Earth]. Notice that it hits the lower half of Earth [cymbals sound as equator glows; lower half turns yellow], called the southern hemisphere, most directly

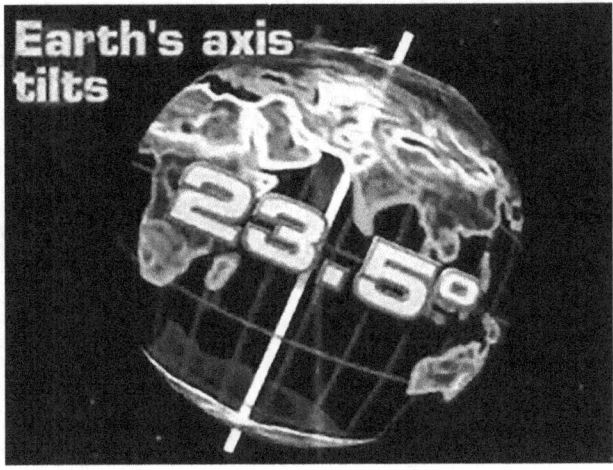

FIGURE 4.6   Seasons animation: Showing Earth's tilt (at 18 seconds)

[hemispheres labels appear]. The Sun's energy hits the northern hemisphere too [sunlight hitting top half turns blue], but look at how the northern hemisphere is pointed away from the Sun [arrow upwards through Earth's axis]. Also, the light hits at an angle [two arrows appear in sunlight band] that causes the energy to be spread out over a greater area. So, at this spot in Earth's revolution it receives less of the Sun's energy.

The narration then describes the effects of different sunlight concentration on hemispheres:

FIGURE 4.7   Seasons animation: Different angles of sunlight hitting the Earth (at 51 seconds)

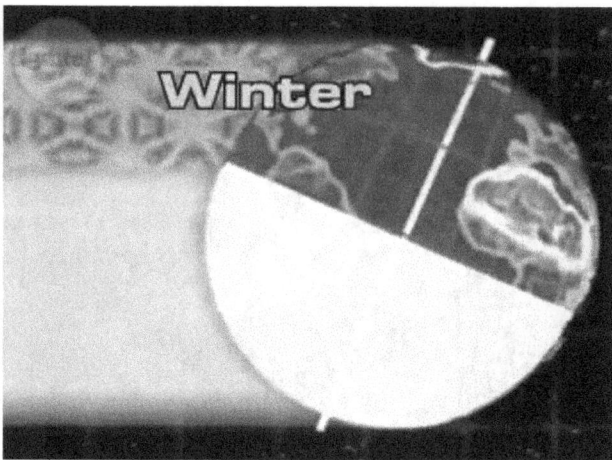

FIGURE 4.8 Seasons animation: Effects of angle and concentration of sunlight – winter (at 1:23 minutes)

When more of the Sun's energy hits the southern hemisphere it causes the temperature to go up [sun patterns and 'Summer' appear in yellow band]. It's summertime! [Sounds of children playing] More sunlight, longer days! With less of the Sun's energy hitting the northern hemisphere, it gets a lot colder [snowflake patterns and 'Winter' appear in blue band; see Figure 4.8]. Put on your winter coat! [Sounds of wind howling] Less sunlight, shorter days!

Third, the animation shows the seasons through the year: 'Let's see how the seasons change as Earth revolves around the Sun'. The caption 'December' appears (to door sound) and the animation proceeds through months of the year (signalled by a keyboard tap sound) as the Earth orbits the Sun. At 'March' the Earth is labelled with 'Spring' with green patterns on the top half and 'Autumn' and brown patterns on the bottom half; at 'June', as illustrated by Figure 4.9, 'Summer' and 'Winter' with yellow and blue patterns; and at 'September', 'Autumn' and 'Spring' with brown and green patterns. Month captions move around the screen, with the door sound when they stop.

Finally, the voice sums up: 'So why do seasons change? Because the tilt of Earth's axis causes the hemispheres to receive different amounts of the Sun's energy'. The Earth orbits the Sun as month captions are tapped through again. The door sound marks the appearance of a photograph of a green hillside next to text stating 'Seasons occur because of Earth's tilt!', as the voice says breathily: 'That's some *cool* science!'.

Affordances for the task

The animation offers limited interactive affordances. It is accessible, shareable, watchable, viewable, listenable, and can be played back. One can choose to only listen or only watch, but the animation is not easily manipulated; for example,

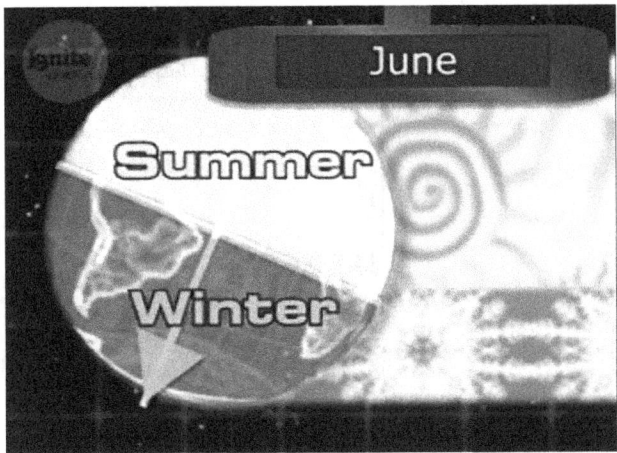

FIGURE 4.9   Seasons animation: Summer/Winter (at 1:41 minutes)

the voice cannot be altered, specific audio elements (such as the music) cannot be turned off and specific content (such as discussion of particular factors) cannot easily be removed. Nonetheless, these affordances for 'instructive' interactions can meet the teacher's purpose of students making notes of key issues explaining the seasons.

Epistemic affordances

The animation's complex mix of elements exhibit different degrees of match to the task. So we shall discuss visual and audio elements in turn, using the translation device of Table 4.4.

Visual elements include text (title, labels, captions), entities (e.g. Earth), movement (e.g. rotating), a photograph and various effects (e.g. yellow band for sunlight). Figure 4.10 locates these elements on the autonomy plane: most are at the far top (PA++) and/or far right (RA++). Overall, the animation thereby visually matches the teacher's core target content and/or purpose. Taking each kind of visual element in turn, text appears frequently, though each briefly. All text – summarized in Figure 4.10 for reasons of space as 'all captions', 'all labels' and 'titles' – is situated in the teacher's core target (PA++, RA++). Entities and movements, such as Earth and its orbit, are also largely in her core target. The exceptions are: Earth's rotation and the Moon and its orbit. Both, however, were previously discussed in the class: Earth's rotation was a factor in an explanation discussed earlier in the lesson (PA+, RA+); and the Moon and its orbit was discussed in the previous lesson (PA–, RA–). Other visual elements are in the teacher's core target: axis line, sunlight band and sunlight arrows are part of the explanation and students' ability to use these symbols in diagrams is checked by the teacher later in the lesson. Other visual effects are mostly brief and located at the bottom right of Figure 4.10. While their contents – colours, patterns, glowing – are far from the seasons (and so coded as unassociated non-target content or PA– –), they all serve the teacher's core target purpose (RA++) by

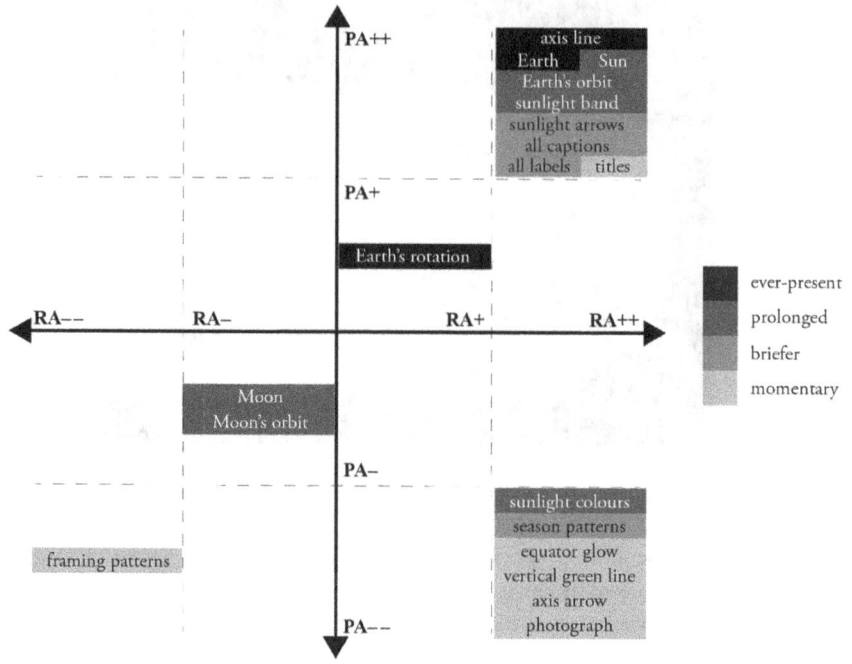

FIGURE 4.10 Visual elements of the seasons animation

emphasizing factors such as different hemispheres and seasons. The one exception is only momentary: patterns framing the static images at the start and end (PA– –, RA– –).

Audio elements include a spoken narration, voice tone, music and various sound effects (e.g. howling wind). The almost ever-present narration, shown as non-italics in Figure 4.11, covers all factors and relations that comprise the explanation of seasons in the teacher's core target (PA++, RA++).[15] It would thus appear ideally suited to the task. However, the narration also includes two further factors: daylight length is discussed in another explanation during the lesson but not in this task (PA+, RA+); and concentration of sunlight, though related to the seasons, is not addressed in the lesson (PA–, RA–). The narration also exudes an excitable, breathy tone, especially in playful phrases (e.g. 'Put on your winter coat!') – these are located at the bottom left of the plane, far from the teacher's core target (PA– –, RA– –). Other audio effects (*italics* in Figure 4.11) are frequent but individually brief. Sounds of rumbling, children playing, etc. represent non-target content (PA– –) but serve the teacher's core target purpose (RA++) by emphasizing factors such as solar energy and seasons. Exceptions are: whirring for Earth's rotation (PA– –, RA+); and the recurrent door shutting sound and ever-present background music (PA– –, RA– –).

In summary, the animation's interactive affordances can support the teacher's 'instructive' task of students making notes of key issues. The animation also exhibits

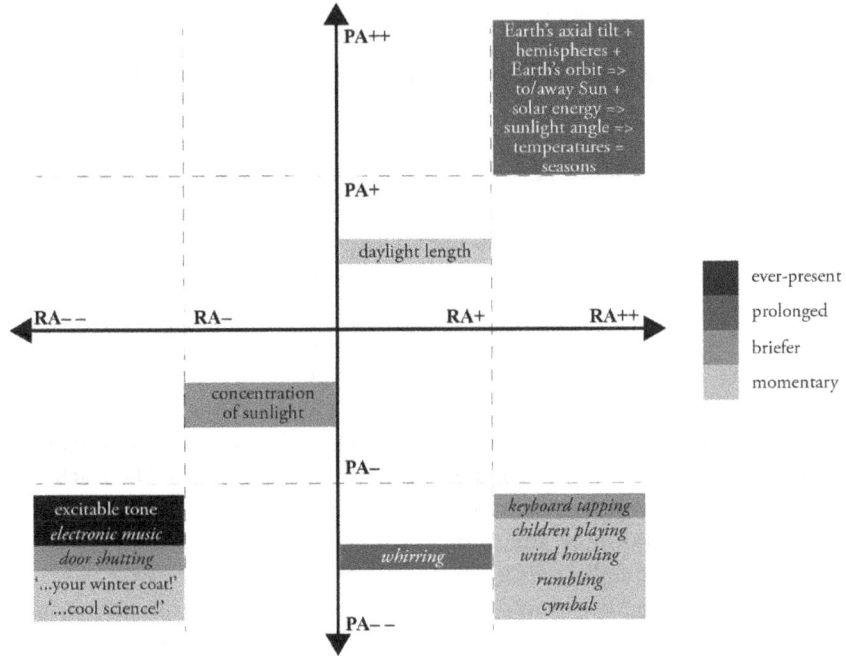

FIGURE 4.11 Audio elements of the seasons animation

significant epistemic affordances for the task. Visually, what does not lie within the teacher's core target is either part of the lesson and unit of study or only a momentary feature of the animation. Aurally, the spoken narration includes all the factors and relations that the teacher wishes to convey. In short, the animation is well selected for both interactive and epistemic affordances. However, as we emphasize, multimedia objects are unlikely to perfectly match a specific task in a specific classroom. In this case, two issues could potentially require pedagogic work. First, factors in the narration that the teacher does not wish to include (daylight length and concentration of sunlight) are not easily subtracted. Second, the ever-present voice tone and music, if not wanted, can only be subtracted by omitting all audio elements. Given that the visual elements do not by themselves provide an explanation, this requires adding a commentary. The success of such *substitution* – replacing non-target with core target elements – would depend on the teacher's ability to discuss core target knowledge while selecting and turning to purpose the animation's visual elements.

Seasons in the Sun

The teacher undertakes the pedagogic work required to maximize the animation's epistemic affordances for supporting her task. She does so by taking advantage of its watchability and listenability, muting the animation and substituting the audio with her own narration over the visual elements. In an interview, she highlighted its terminology, music and voice tone as reasons – epistemic rather than interactive affordances:

> I didn't use the sound in that one because I wanted to be able to use the language that I was using in class, that the sound in that animation was a very hard American voice, almost computerised. There was this tinkly music in the background. I thought, 'This is going to be distracting'.

As we state above, such substitution depends on a teacher's ability to add audio content while activating an animation's visual elements. In this case, the teacher is highly adept. Without pausing the animation, she highlights, through her speech or gestures, specific entities, movements, visual effects and text while introducing factors and explaining how they relate together to create the seasons. The teacher also asks questions of students that are timed so that the appearance of captions confirms their responses. In short, the animation and teacher narration form an integrated multimedia and multimodal performance.

To show the pedagogic work this involves, we shall discuss the task using the same four stages as the animation's voiceover (above). As outlined earlier, her core target is to explain how 'the tilt of our axes and our position around the Sun' shape Earth's seasons. First, she highlights the key issue of tilt:

> So [reading out title] what causes Earth's seasons? When we have a little look at this animation, what we're looking at here is obviously the Earth spinning on its axis and you can see that that axis [hand gesture matching its axis line] is about 23.5 degrees [points to axis line] from [hand traces vertical line] what could be the [air quotes gesture] "theoretical midline" of the Earth. We know [repeatedly points to Earth as it orbits Sun] that that axis holds itself. Can you see [pointing to Earth] how it's holding its 23.5 degrees as it moves around the Sun?

Figure 4.12 plots key elements of her narration as a whole.[16] As the above shows, she begins by quickly turns from describing what is being shown (Earth rotates) to the issue of axial tilt, selecting visual elements relevant to her core target: Earth's axis, tilt and orbit (PA++, RA++). The teacher's regular gestures are non-target content (PA– –), because hand movements are far from her target content, but they serve her core target purpose (RA++) by, for example, highlighting tilt.

Second, the teacher set out an explanation of how tilt leads to differences in the angle of sunlight in different hemispheres:

> Now, [pointing to yellow band from Sun] when the Sun's rays hit the Earth, the Earth has that also other [air quotes] "theoretical midline", the equator [hand traces equator line], that breaks it into half: [pointing to labels] northern hemisphere and southern hemisphere. Okay? We're [pointing to herself] in the

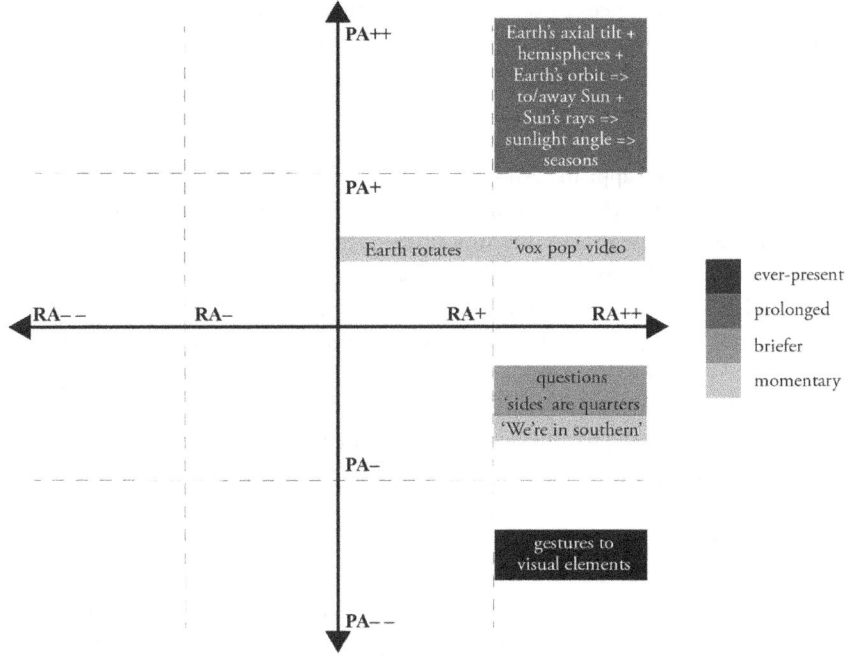

FIGURE 4.12 Teacher narration of the seasons animation

southern hemisphere and at some times [tracing horizontal line in yellow band] the sunlight will strike, in this case, the southern hemisphere at that 90 degree angle. All right? And here [hand tracing angled arrow in sunlight band], this is what we can sometimes call an 'oblique'. You heard that person in the vox pop talk about it. Yeah? Or 'glancing'. So [hand tracing where angled arrow had been] these are like the 15 degrees, okay? We're getting sunlight bouncing off, and here [hand tracing where horizontal arrow had been] it's hitting directly.

Through this explanation the teacher's gestures continue linking visual elements to her core target knowledge: the sunlight band is recruited into discussing the Sun's rays, the equator line helps identify hemispheres, and the arrows are mirrored by gestures that indicate sunlight angles. In comparison to the muted narration, she offers a simpler explanation by not discussing the effects of sunlight angle on daylight length or on sunlight concentration and their effects, in turn, on temperatures. Instead, these elements are substituted by: locating herself and her students in the southern hemisphere; providing more information about sunlight angles, specifically the terms 'oblique' and 'glancing'; and explicitly connecting that information to a 'vox pop' video they watched earlier in the lesson. Everything she adds serves her core target purpose (far right on the Figure, or RA++) and helps integrate the task and animation into the rest of the lesson and thus the classroom experiences shared by students.

Third, the teacher checks student understanding of the effects of sunlight angle on seasons and, emphasizing the Earth's tilt during its orbit of the Sun, discusses how seasons change through the year:

> So, in this case, what season does the southern hemisphere have? [Students say 'summer', as label appears]. And what about the northern hemisphere? [Students say 'winter']. Winter! Good. Now, when this moves [gesturing to Earth] around the Sun, and holds its axis, because it doesn't … the axis won't change around like that, when this moves to the other side of the Sun, we will see the other side of the Earth … so [pointing to screen showing Earth with 'spring' and 'autumn' labels] now we're on the back side of the Sun. We've got spring and autumn. Now we're on the opposite side [Earth with 'summer' and 'winter' labels]. Now the northern hemisphere has summer and the southern hemisphere has winter. Now, we come out to the [air quotes] "fourth side" of the Sun [Earth with 'autumn' and 'spring' labels], we get autumn and spring split around.

Here the questions to students are not part of the explanation (PA−) but turned to her core target purpose (RA++), as shown on Figure 4.12. She then recruits visual entities (the Earth and the Sun), movements (Earth orbiting) and labels (e.g. 'winter') to discuss the changing seasons, remaining throughout inside her core target.

Finally, the animation continues as a student queries the description of 'sides' and the teacher explains that 'sides' is a way of emphasizing the change of seasons by referring to quarters of Earth's orbit: non-target content turned to her core target purpose (PA−, RA++). The animation's final textual summary is not referred to by the teacher. Instead, after it concludes she sums up: 'So, we know the Sun is hitting the Earth. We know that some parts of the Earth will be getting the full force and some will be getting those glancing rays.' She asks the students whether the Sun changes and emphasizes: 'typically the strength of the Sun never changes. The thing that will change is our position in relationship to the Sun … We get seasons because of being tilted towards or away'. This concludes the task; the teacher then moves to a task involving an animated simulation that shows the seasons change through the year.

In summary, the teacher uses the animation's ability to be watched and muted in order to reshape its epistemic affordances. Her substitution of audio elements with her own narration and gestures adds elements that are almost entirely on the far right of Figure 4.12 – everything she says or does is turned to her core target purpose. Indeed, Figure 4.12 underplays the extent to which her performance as a whole, and not just her audio narration, is replete with core target knowledge because, for reasons of space, we summarized 'gestures to visual elements' rather than included all those elements she activates in this way. Returning to the plane of visual elements in Figure 4.10, she draws extensively on entities, movements and labels located within her core target: axis line, Earth, Sun, orbit, sunlight band, arrows, titles, captions and labels. In contrast, she ignores those visual elements outside her target. Those which could have served her purpose – the visual effects at the bottom right of Figure 4.10 – were substituted by hand gestures bringing attention to key factors. So, the teacher selects the animation's visual elements that match her task and integrates those elements through gesture with an audio narration that provides a simpler, more focused explanation that is, in addition, more integrated into the wider lesson and classroom experience. Her strategy of substitution thereby not only activates the epistemic affordances of the animation but more closely integrates the multimedia object into the task and lesson.

Conclusion

Multimedia objects such as animations are a common part of teaching science. Questions of how teachers can integrate such ready-made objects into classroom tasks are urgent. However, such questions are sidelined by the cognitive–learning–design axis dominating research into multimedia learning and technology affordances. To address questions of integration requires studies of actual classroom practice and a framework that does not reduce teaching to designing and what is being taught and learned to homogeneous information. In this chapter we extended the metaphor of 'affordances' beyond its current focus on enabling students interactions to embrace the *epistemic affordances* of knowledge practices. We drew on the LCT dimension of Autonomy to bring one aspect of these knowledge practices into view. The concepts of *positional autonomy* and *relational autonomy* helped foreground relations between the contents and purposes of multimedia and those of classroom tasks. The concept of *targets* helped ensure that specific classroom practices remained the centre of analysis, thereby avoiding the slide into generic descriptions of context-free 'abilities' towards which affordance frameworks tend. These concepts provided a platform for seeing that teaching is more than design, that integration is more than selection, and that affordances are more than enabling interactive practices among students.

We illustrated the value of seeing epistemic affordances with two examples of teaching science with animations. Both teachers selected animations whose limited interactive affordances matched their 'instructive' needs – from the conventional perspective of 'affordances' there is little to distinguish between them. Yet they offer contrasting examples of teaching science with multimedia. The first teacher did not undertake the pedagogic work required to activate the epistemic affordances of the Moon animation for supporting his task (teaching about synchronous rotation) and left that task segmented from the rest of the lesson. The second teacher undertook pedagogic work by substituting the audio elements in the seasons animation that did not match her specific task with her own narration and connecting this knowledge to its visual elements through gestures. From the viewpoint of epistemic affordances for teaching scientific explanations, then, these examples are radically different.

By bringing teaching and knowledge practices into view, autonomy analysis emphasizes that integration does not end with design or selection. While multimedia are unlikely to perfectly match the knowledge needs of a specific classroom task, that is not necessarily an obstacle to their integration. Elements beyond a teacher's core target offer possibilities for teaching and pedagogic work by teachers and students can 'connect to content' (strengthen PA) and 'turn to purpose' (strengthen RA) those elements of a resource that lie beyond their core targets. Thus, ready-made multimedia objects can be integrated through classroom practice. Moreover, the epistemic affordances of a multimedia object may differ wildly from one task to the next. The coding of an object is in relation to targets that are highly specific: a particular task in a particular lesson. The same multimedia objects analyzed in relation to other teacher's targets, to other tasks of these teachers or to students' targets will result in different codings. Epistemic affordances depend on the specific knowledge practices being taught and learned. Thus, design and selection are not the end of the matter: integration is a practice by teachers and students in classrooms.

Of course, seeing teaching, knowledge practices and epistemic affordances is but a small step forward. One cannot expect teachers to undertake time-consuming autonomy analyses of tasks and multimedia. The next step requires more practice-oriented outcomes. Autonomy analyses of classroom practices that integrate mathematics into science (Maton and Howard, chapter 2, this volume) and everyday experiences into History teaching (Maton and Howard 2018) have developed the pedagogic practice of 'autonomy tours', which is already having an impact on teaching practice. Similar pedagogic principles are required for integrating multimedia into teaching. Nonetheless, this chapter has illustrated that LCT offers the potential for addressing such issues of integration. All we have to do now is to activate those affordances.

Notes

1 This holds for models of both 'technological' affordances and also 'social' and 'educational' affordances (Kirschner *et al.* 2004) – all concern interactions.
2 We use 'epistemic affordances' to encompass all knowledge practices, whether involving epistemological constellations of concepts and empirical descriptions or axiological constellations of affective, moral, ethical or political meanings (Maton 2014: 148–70).
3 See: www.legitimationcodetheory.com.
4 There is a growing body of work using Semantics to study multimedia in science, such as student-generated digital products (Georgiou 2020, Georgiou and Nielsen 2020). From a complementary perspective, He (2020) explores the semiotic resources of science animations with systemic functional linguistics.
5 Students, other teachers, etc. may have different targets, allowing comparative analysis of, for example, learning experiences.
6 We draw on research funded by the Australian Research Council (DP130100481), led by Karl Maton, J. R. Martin, Len Unsworth and Sarah K. Howard.
7 The concepts can be enacted in different ways. In Maton and Howard (2018, chapter 2, this volume), we focus on changes in knowledge practices over time and so plot 'autonomy pathways' between different 'autonomy codes' on the plane. Here we do not name autonomy codes, as our focus is on 16 rather than four modalities, and we plot a synchronic analysis of positions, as our focus is relating task and object rather than changes in strengths through time.

8 The animation is freely available on the Internet. The earliest we discovered was uploaded to YouTube in 2009 by 'astrogirlwest', about whom no further information is available.
9 For both animations analysed in this chapter, it is easy to imagine different resources that could make the tasks more collaborative for students. However, that would not diminish their ability to support 'instructive' interactions and thus match the teachers' core target purposes.
10 Relative duration is, of course, but one indicator of prominence – one could also examine relative size, position, and other attributes.
11 One uploader to YouTube added their own starting caption: 'This is a video explaining why we always see the same side of the moon all the time'; see https://www.youtube.com/watch?v=6vkVxu04DcE.
12 The teacher confuses the duration between New Moons (29.5 days) with the Moon completing a revolution on its axis (27.3 days). However, accuracy here would not change the degree to which he integrates the animation.
13 These factors are highlighted in a constellation analysis of secondary school science textbooks by Maton and Doran (chapter 3, this volume).
14 The animation is dated 2006, accredited to Ignite! Learning, and freely available at https://www.teachertube.com/videos/what-causes-seasons-on-earth-657.
15 We have included the logic of its explanation by using '+' and '=>' to refer to links between nodes; see Maton and Doran (chapter 3, this volume).
16 We have highlighted the logic of her explanation with '+' and '=>'. Maton and Doran (chapter 3, this volume) discuss this logic in greater detail.

References

Antonenko, P., Dawson, K. and Sahay, S. (2017) 'A framework for aligning needs, abilities and affordances to inform design and practice of educational technologies', *British Journal of Educational Technology*, 48(4): 916–27.
Berney, S., and Bétrancourt, M. (2016) 'Does animation enhance learning?: A meta-analysis', *Computers & Education*, 101: 150–67.
Bower, M. (2008) 'Affordance analysis – matching learning tasks with learning technologies', *Educational Media International*, 45(1): 3–15.
Bower, M. (2017) *Design of Technology-Enhanced Learning: Integrating Research and Practice*, Bingley: Emerald Publishing.
Georgiou, H. (2020) 'Characterising communication of scientific concepts in student-generated digital products', *Education Sciences*, 10(1): https://doi.org/10.3390/educsci10010018.
Georgiou, H. and Nielsen, W. (2020) 'New assessment forms in higher education: A study of student generated digital media products in the health sciences', in C. Winberg, S. McKenna and K. Wilmot (eds) *Building Knowledge in Higher Education*, London: Routledge, 55–75.
Hammond, M. (2010) 'What is an affordance and can it help us understand the use of ICT in education?', *Education and Information Technologies*, 15(3): 205–17.
He, Y. (2020) '*Animation as a Semiotic Mode: Construing Knowledge in Science Animated Videos*', unpublished PhD thesis, University of Sydney, Sydney.
Howard, S. K. and Maton, K. (2011) 'Theorising knowledge practices: A missing piece of the educational technology puzzle', *Research in Learning Technology*, 19(3): 191–206.
Howard, S. K. and Maton, K. (2013) '*Technology & knowledge*', *AERA 2013 SIG-Computer and Internet Applications in Education*, San Francisco, USA: AERA, 1–8.

Jenkinson, J. (2018) 'Molecular biology meets the learning sciences: Visualizations in education and outreach', *Journal of Molecular Biology*, 430(21): 4013–27.

Kirschner, P., Strijbos, J.-W., Kreijns, K. and Beers, P. J. (2004) 'Designing electronic collaborative learning environments', *Educational Technology Research & Development*, 52(3): 47–66.

Li, J., Antonenko, P. D. and Wang, J. (2019) 'Trends and issues in multimedia learning research in 1996–2016: A bibliometric analysis', *Educational Research Review, 28*; 100282

Maton, K. (2014) *Knowledge and Knowers: Towards a Realist Sociology of Education*, London: Routledge.

Maton, K. (2020) 'Semantic waves: Context, complexity and academic discourse', in J. R. Martin, K. Maton and Y. J. Doran (eds) *Accessing Academic Discourse,* London, Routledge, 59–85.

Maton, K. and Chen, R. T-H. (2016) 'LCT in qualitative research: Creating a translation device for studying constructivist pedagogy', in K. Maton, S. Hood and S. Shay (eds) *Knowledge-building: Educational studies in Legitimation Code Theory*, London: Routledge, 27–48.

Maton, K. and Howard, S. K. (2018) 'Taking autonomy tours: A key to integrative knowledge-building', *LCT Centre Occasional Paper* 1: 1–35.

Mayer, R. E. (2003) 'The promise of multimedia learning: Using the same instructional design methods across different media', *Learning and Instruction*, 13(2): 125–39.

Mayer, R. E. (2014a) 'Cognitive theory of multimedia learning', in R. E. Mayer (Ed.) *The Cambridge Handbook of Multimedia Learning*, 2nd edition, Cambridge: Cambridge University Press, 43–71.

Mayer, R. E. (2014b) (Ed.) *The Cambridge Handbook of Multimedia Learning*, 2nd edition, Cambridge: Cambridge University Press.

Mayer, R. E. and Fiorella, L. (2014) 'Principles for reducing extraneous processing in multimedia learning', in R. E. Mayer (Ed.) *The Cambridge Handbook of Multimedia Learning*, 2nd edition, Cambridge: Cambridge University Press, 279–315.

Mayer R. E. and Moreno, R. (2003) 'Nine ways to reduce cognitive load in multimedia learning', *Educational Psychologist*, 38(1): 43–52.

Mutlu-Bayraktar, D., Cosgun, V. and Altan, T. (2019) 'Cognitive load in multimedia learning environments: A systematic review', *Computers & Education*, 141: 103618.

Ploetzner, R. and Lowe, R. (2012) 'A systematic characterisation of expository animations', *Computers in Human Behaviour*, 28: 781–94.

Sweller, J., van Merriënboer, J. and Paas, F. (2019) 'Cognitive architecture and instructional design: 20 years later', *Educational Psychology Review, 31*: 261–92.

PART II

Language in science education

5

FIELD RELATIONS

Understanding scientific explanations

Y. J. Doran and J. R. Martin

Making sense

If you live away from the equator, the seasons are clear. Summer is hot, winter is cold and spring and autumn are somewhere in between, harbingers of what's to come. If you are a student of secondary-school science, this often becomes the starting point for a considerably more academic discussion: why do seasons occur? And the resulting explanation is far removed from everyday experience. It involves the rotation of the earth, its tilt, its division into northern and southern hemispheres, its orbit around the sun, the sun's emission of light and a number of effects resulting from some combination of these. Our everyday understanding of hot and cold throughout the year quickly transforms into a large complex of scientific 'facts'.

An explanation oriented to a popular audience, for example, explains: [1]

What Causes Seasons on Earth?

Seasons happen because Earth's axis is tilted at an angle of about 23.4 degrees and different parts of Earth receive more solar energy than others.

Because of Earth's axial tilt (obliquity), our planet orbits the Sun on a slant which means different areas of Earth point toward or away from the Sun at different times of the year.

Around the June solstice, the North Pole is tilted toward the Sun and the Northern Hemisphere gets more of the Sun's direct rays. This is why June, July and August are summer months in the Northern Hemisphere.

Opposite Seasons

At the same time, the Southern Hemisphere points away from the Sun, creating winter during the months of June, July and August. Summer in the Southern Hemisphere is in December, January, and February, when the South Pole is tilted toward the Sun and the Northern Hemisphere is tilted away.

Such an explanation pushes us far away from our everyday experience. But for students to be successful in science, they must be able to both read and write explanations of this kind regularly. Indeed explaining why seasons occur is a key topic in the first year of secondary school science in New South Wales, Australia.

In recent years, dialogue between Systemic Functional Linguistics (SFL) and Legitimation Code Theory (LCT) has been exploring the nature of such knowledge and the discourse used to organize it (Martin 2011, Maton *et al.* 2016, Maton and Doran 2017, Martin *et al.* 2020). A goal of this research across disciplines is to develop discipline-sensitive pedagogy and curriculum to improve the learning of subject-specific knowledge. As part of this endeavour, research has focused on different kinds of knowledge, how knowledge develops over time and how we can model the knowledge we find in spoken and written texts and across the vast range of multimodal semiotic resources used in school.

In terms of Legitimation Code Theory, for students to master the scientific knowledge involved in explaining the seasons, they need to be able to understand and manipulate intricate *constellations* of meaning (Maton 2014: 149–159; Maton and Doran, Chapter 3 of this volume). These constellations organize large networks of meaning into specific arrangements according to the discipline, the year level and the phenomenon being explained. In science, what is foregrounded are what LCT terms *epistemological constellations* which focus on the 'content' of disciplines – discipline-specific configurations of causal relations, taxonomies and scientific procedures, along with methods for investigating the world. This is in contrast to *axiological constellations* that emphasize specific dispositions, political and aesthetic stances, morals and ethics, often found for example in the humanities (Martin *et al.* 2010, Maton 2014, Doran 2020).

From an epistemological constellation perspective, the key components in the constellation underpinning why seasons occur in this explanation can be synthesized from the text as follows:

Earth's axial tilt is 23.4 degrees
The amount of solar energy received by different parts of the earth at different times of the year varies.
The earth orbits around the sun
The earth is divided into hemispheres
Seasons occur

These components are not simply a bunch of isolated 'facts' about seasons. In order to explain seasons, they need to be brought together in specific configurations. For example, the variation in the amount of solar energy received by different parts of the earth that underpins seasons is caused by the particular combination of the Earth's 23.4 degree tilt, its orbit around the sun, its division into hemispheres and the fact that the sun emits light that hits the earth. Without any of these components, seasons would not occur.

The questions then for this chapter are: what are the relations that underpin scientific knowledge? And how do we *see* these relations in texts, both through language and other semiotic resources? Put another way, we are concerned with what it is that students are expected to learn when they 'do' science.

As far as SFL is concerned, one vantage point from which to explore this is through the register variable *field*. In SFL, field is one component of register, along with tenor and mode, and is concerned with what educators consider the content of language and semiosis. Seen in terms of SFL, field is a more abstract level of meaning positioned above the ideational meanings construed through language at the levels of discourse semantics, lexicogrammar and phonology/graphology. These levels of abstraction, called strata, are represented as co-tangential circles in Figure 5.1 (Martin 1992).

Over a number of years, work on field and ideational meaning has provided a highly productive lens through which to view scientific understandings of the world (e.g. Lemke 1990, Halliday and Martin 1993, Martin and Veel 1998, Halliday 2004, Martin 2020). More recently, the model of field has been renovated, largely in response to SFL's dialogue with LCT (Martin and Maton 2013, Martin *et al.* 2017, Martin *et al.* 2020), the development of ideational discourse semantics by Hao (2015, 2018, 2020a, 2020b, 2020c, Chapter 6 of this volume), and investigations of a range of semiotic resources used in science, including mathematics, graphs, diagrams, animations, other formalisms and body language (Doran 2017, 2018, 2019, Chapter 7 of this volume; Unsworth 2020; He 2020; Hood and Hao, Chapter 10 of this volume; Martin *et al.* in press).

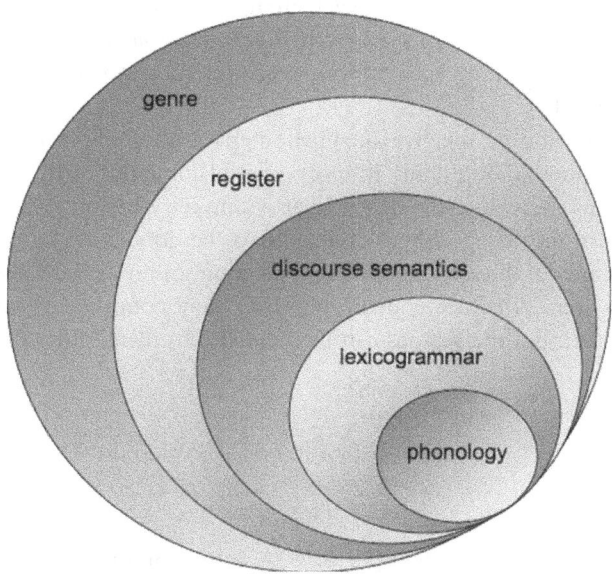

FIGURE 5.1 Strata of language in Systemic Functional Linguistics

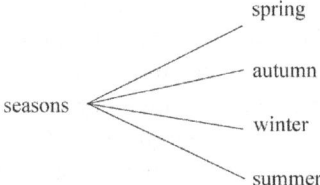

FIGURE 5.2 Classification taxonomy of the seasons

This paper outlines these evolving understandings. It will introduce each component of the new model, drawing on a range of linguistic texts and semiotic resources. It will then use the model to explore seasons as they are taught in secondary school in New South Wales, Australia. In doing so, the chapter illustrates how field can make visible the knowledge teachers need to teach and students need to learn in order to achieve success in school science.

Field relations: static and dynamic perspectives

The model developed here expands on that of Martin (1992). At its most general level, field can be described as a resource for construing phenomena either *statically* as relations among items or *dynamically* as activities oriented to some global institutional purpose. Beginning with the *static* perspective, this orientation to field views phenomena as items organized into particular taxonomies. For example, our common-sense understanding distinguishes four seasons: winter, summer, autumn and spring. This can be modelled as a classification taxonomy, as in Figure 5.2.

Classification views relations between items in terms of types and sub-types (class and sub-class). In terms of seasons, the items *spring, autumn, winter* and *summer* are all sub-classes in relation to the more general item of *seasons* and co-classes in relation to each other.

An alternate static perspective on phenomena is through composition – the part-whole relations among items. In terms of explaining the earth's seasons, composition relations are used to divide the solar system into the sun and the earth, and the earth in turn into the northern and southern hemispheres. The relevant composition taxonomy of the solar system for explaining seasons is shown in Figure 5.3.

Our solar system of course contains many different components aside from the earth and the sun, and the seasons can be divided into many other configurations

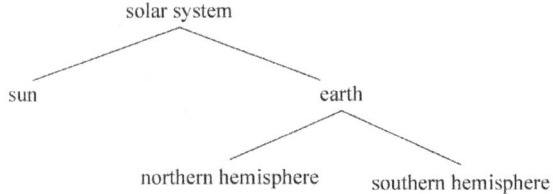

FIGURE 5.3 Composition taxonomy of the solar system

(northern Australians may well wonder where the wet season or the dry season or the build-up are in the classification introduced above). But the important point here is that for the particular problem of explaining how seasons work in junior high school, ethnocentric though it may be, these are the composition and classification relations that are relevant. In LCT terms, these meanings form part of the specific *constellation* comprising knowledge of the seasons.

As far as our model of field is concerned, taxonomies may be indefinitely wide or deep. For example, the composition taxonomy introduced above indicates two parts for each level, whereas the classification taxonomy distinguishes four seasons. And in principle, there may be any number of subtypes or parts depending on what one is looking at. Similarly, the composition taxonomy of the solar system shows a slightly deeper hierarchical arrangement than the classification taxonomy, with three levels of the composition (solar system → earth → northern hemisphere, for example). Again, this may be expanded indefinitely, with any number of parts and wholes relevant for a given field. Indeed the expansion of taxonomies is one of the key features of scientific knowledge. This is one respect in which science differs from common-sense understandings whose utility tends to demand less width and depth of classification and composition (Wignell *et al.* 1989).

We can begin mapping the types of relations underpinning fields with the network in Figure 5.4. This network says that from a static perspective, a field may involve a single item or multiple items arranged into taxonomies of either composition or classification. The *n* above taxonomy indicates it may be indefinitely wide and/or deep.

An alternate perspective on field involves construing phenomena *dynamically* in terms of activities. Activity involves some sort of change that is oriented to some global everyday, professional or institutional purpose. One activity associated with the seasons, for example, is:

The sun warms the earth

This example gives a single, isolated activity, specified lexicogrammatically by the Process *warms*, and involving two items: *the sun* and *the earth*. In longer explanations, rather than construing just a single undivided activity, it is common for an activity to be construed as a series of smaller activities. <u>For example, the above activity could be reconstrued as a series of activities as follows:</u>

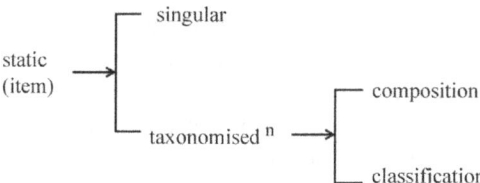

FIGURE 5.4 Network for a static perspective on field

The sun emits light

∧

which hits the earth

∧

and warms it.

Here we have a *momented activity*, where a single activity (*the sun warms the earth*) is construed as three separate activities, all oriented to the more general activity of warming.[2]

Each activity can be momented into any number of further activities within the limits of a field. For example, in Figure 5.5 below, from a secondary school biology textbook (Greenwood and Allen 2004), the unmomented activity *inflammation* is momented through various subheadings into four stages:

Increased diameter and permeability of blood vessels

∧

Phagocyte migration

∧

Phagocytosis

∧

Tissue repair

Each of these activities can in turn be momented into another series of activities. For example, at the bottom of the figure, *phagocytosis* is momented as:

Detection

∧

Ingestion

∧

Phagosome forms

∧

Fusion with lysosome

∧

Digestion

∧

Discharge

This produces three tiers of activity shown in Figure 5.5.

An unfolding series of activities can be related in one of two ways, through *implication* or *expectancy*. Implication relations describe series of activities where one activity necessarily entails another (if one activity happens, then another always does).[3]

In the following momented activity from a year seven science classroom, the activity *tides* is described as a necessary result of the *gravitational pull of the moon and the sun* and *the rotation of the earth*, through the Process *causes*:

Inflammation

Tier 1

Damage to the body's tissues can be caused by physical agents (e.g. sharp objects, heat, radiant energy, or electricity), microbial infection, or chemical agents (e.g. gases, acids and bases). The damage triggers a defensive response called **inflammation**. It is usually characterised by four symptoms: pain, redness, heat and swelling. The inflammatory response is beneficial and has the following functions: (1) to destroy the cause of the infection and remove it and its products from the body; (2) if this fails, to limit the effects on the body by confining the infection to a small area; (3) replacing or repairing tissue damaged by the infection. The process of inflammation can be divided into three distinct phases. These are detailed below.

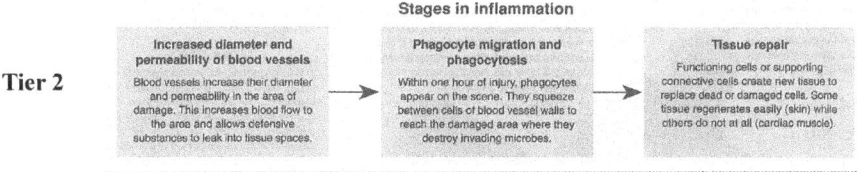

Tier 2

Stages in inflammation

Increased diameter and permeability of blood vessels
Blood vessels increase their diameter and permeability in the area of damage. This increases blood flow to the area and allows defensive substances to leak into tissue spaces.

Phagocyte migration and phagocytosis
Within one hour of injury, phagocytes appear on the scene. They squeeze between cells of blood vessel walls to reach the damaged area where they destroy invading microbes.

Tissue repair
Functioning cells or supporting connective cells create new tissue to replace dead or damaged cells. Some tissue regenerates easily (skin) while others do not at all (cardiac muscle).

Tier 3

How a Phagocyte Destroys Microbes

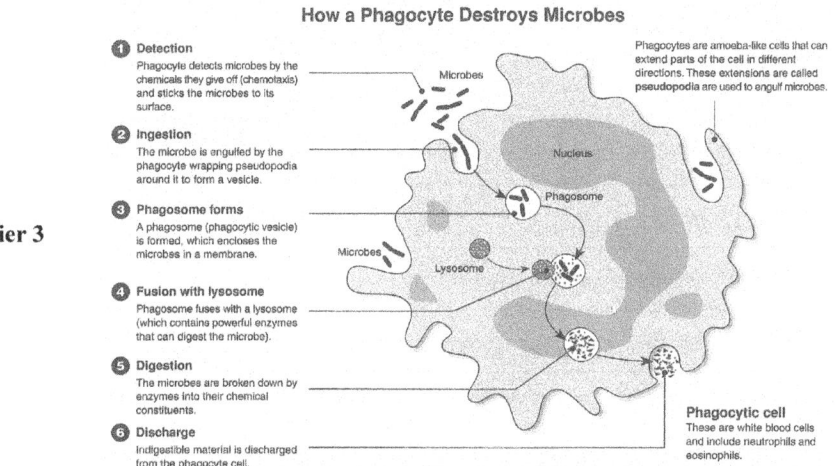

1 Detection
Phagocyte detects microbes by the chemicals they give off (chemotaxis) and sticks the microbes to its surface.

2 Ingestion
The microbe is engulfed by the phagocyte wrapping pseudopodia around it to form a vesicle.

3 Phagosome forms
A phagosome (phagocytic vesicle) is formed, which encloses the microbes in a membrane.

4 Fusion with lysosome
Phagosome fuses with a lysosome (which contains powerful enzymes that can digest the microbe).

5 Digestion
The microbes are broken down by enzymes into their chemical constituents.

6 Discharge
Indigestible material is discharged from the phagocyte cell.

Phagocytes are amoeba-like cells that can extend parts of the cell in different directions. These extensions are called pseudopodia are used to engulf microbes.

Phagocytic cell
These are white blood cells and include neutrophils and eosinophils.

FIGURE 5.5 Momenting the activity *inflammation* (three tiers) (Greenwood and Allen 2004: 118–119)

'The gravitational pull of the moon and the sun, and the rotation of the earth causes tides.'[4]

Implication series are most common in scientific explanations where events tend to be described in terms of causal or conditional relations of entailment. In terms of the ideational discourse semantic model developed by Hao (2020a), the activities *gravitational pull of the moon and the sun* and the *rotation of the earth* are realized by occurrence figures in discourse semantics that are in turn realized metaphorically through nominal groups, while the activity *tides* is realized by an activity entity (see Hao, Chapter 6 of this volume). The implication relation linking these activities is realized discourse semantically by a causal connexion and lexicogrammatically by the Process *cause*.[5] This gives the implication series:

The gravitational pull of the moon and the sun, and the rotation of the earth
^ (causes)
tides

Analyzed for multiple strata, the realization of this momented activity in discourse semantics and lexicogrammar is:

The gravitational pull of the sun and the moon and the rotation of the earth causes tides

field	momented implication activity
disc. sem.	sequence
lexicogram.	clause

Viewed in terms of the realizations of individual activities, this example is:

The gravitational pull of the sun and the moon	*and*	*the rotation of the earth*	*causes*	*tides*
activity		activity		activity
occurrence figure	connexion	occurrence figure	connexion	activity entity[6]
nominal group		nominal group	verbal group	nominal group

If we unpack the grammatical metaphor in this example, we can see the implication series more clearly:

The moon and the sun are pulled by gravity
and the earth rotates,
^ (so)
tides occur.

The use of the consequential connexion (*causes* or *so*) indicates that the tides necessarily arise due to the other two activities. As Hao (2018) has shown, in scientific explanations implication series can also be realized by temporal connexions. For example, both series of activities below are related through implication (where one activity entails another), although one uses conditional connexion (*if... then*) and the other involves temporal connexion (*when*):

If you move an electron from one orbit to another then energy is either absorbed or released.
When you move an electron from one orbit to another, energy is either absorbed or released.

Whereas scientific explanations tend to be concerned with entailment, texts orienting toward how to *do* science (such as experimental procedures, procedural recounts or protocols, or stories of scientific discoveries) tend to construe activity in terms of expectancies rather than as logical necessities (Hood 2010, Hao 2015, 2020a). In these texts, temporal connexions are not used to realize a series of activities related by implication, but rather by *expectancy*. For example, in the following description from a video shown in a high-school science classroom of how data is collected to measure tides, the *gathering* and *uploading* are expected to occur together in a temporal sequence (linked by *and then*), but one activity does not necessarily entail the other. This kind of text records scientific activity rather than offers a scientific explanation of phenomena.

'The data is gathered from the station and then is uploaded via the satellite every 6 minutes'

One of the key features of expectancy relations is that the series of activities can be interrupted or go against what is expected (as any science teacher or observer of a classroom experiment can attest). An interruption of an expectancy series is often shown through concessive connexion – for example *but* in *The data was gathered from the station but was not uploaded via the satellite*. The ability to play with expectancy in momented activities in this way is the basis of many story genres, where some unexpected and often dramatic activity occurs to establish the main complication of the story (Martin and Rose 2008).

Individual activities can take a number of forms. In the broadest terms, all activities involve some sort of change. As exemplified by the *orbit* and *revolve* activities in the explanation of seasons given above, this change can often be cyclical:

Our planet orbits the sun on a slant.
Earth revolves around the sun

Similarly, in the following brief explanation of tides, *rotation* indicates a cyclical activity:

The gravitational pull of the moon and the sun, and the rotation of the earth causes tides.

These cyclical activities involve an event that recurs an indefinite number of times. Put another way, if *rotation* or *orbit* were to be momented, the cycle would repeat indefinitely. Indeed, this is made explicit in a teacher's description of the rotation of the earth, where the cyclical activity *spinning* is momented into *it's moving day night, day night*:

The earth is spinning on its axis, happening every day. Every time it spins, it's moving day night, day night.

In contrast, many activities involve linear unfolding, without recurrence. Such linear activities may involve some sort of culmination, as in the following when the motion of light ends at the earth:

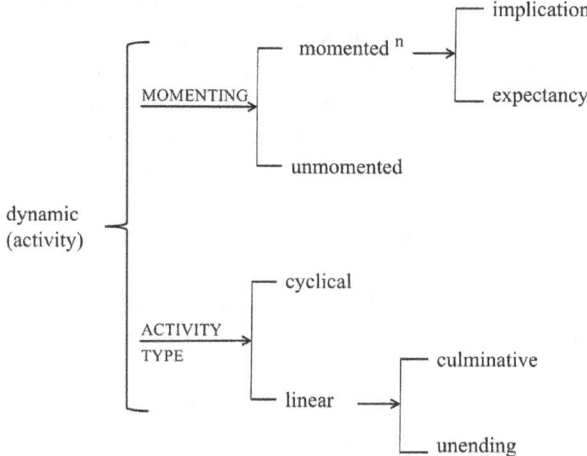

FIGURE 5.6   Network for a dynamic perspective on field

The sun's light hits earth

Alternatively there may be no indication of an end point:

The sun emits light

These options are pulled together into a network in Figure 5.6. This network says that if a dynamic perspective is chosen, there are two sets of options available (the curly bracket { indicates simultaneous choices): if dynamic (activity), then choose from *both* MOMENTING and ACTIVITY TYPE). Within MOMENTING, an activity can be unmomented (given as a single whole), or it can be divided on another tier into multiple activities through the choice of momented. The superscript ^{n} indicates that an activity may be momented an indefinite number of times. If an activity is momented, then the series of activities which moment it may be related through implication or expectancy. The ACTIVITY TYPE system indicates that each activity may be either cyclical or linear, and if linear, may be either culminative or unending.

The dynamic activity and static item options offer alternative but complementary perspectives on phenomena. Although in any particular field one perspective will tend to be emphasized over the other, in principle all phenomena can be viewed in either way. For example, the cardiovascular system within the biological sciences can be viewed statically as a composition taxonomy of constituent items, including the heart, lungs, veins and arteries; or it can be viewed dynamically as the circulation of blood and transportation of nutrients, oxygen and the like to nourish, help fight disease and stabilize temperature. To show these alternate perspectives, the network in Figure 5.7 brings together the options for both a dynamic and static perspective on field.

As far as explanations of seasons are concerned, we can use this network to show how particular sets of activities and taxonomies form its basic building blocks. The

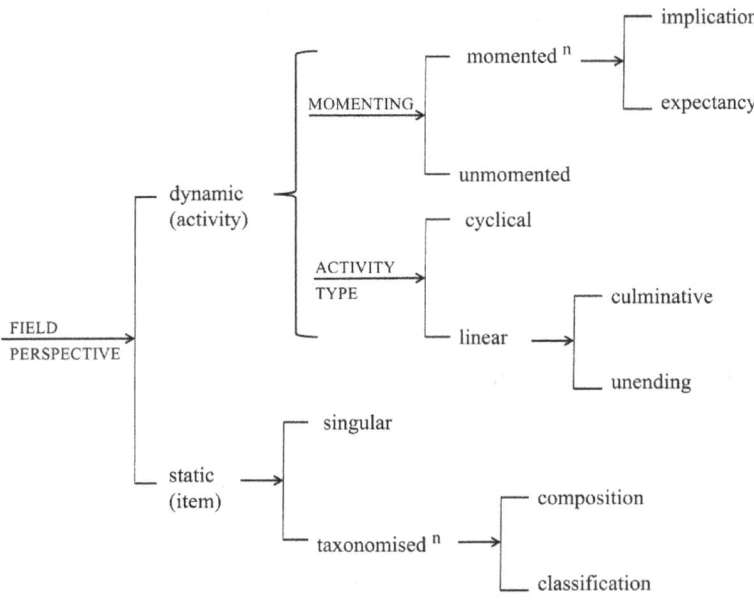

FIGURE 5.7   Dynamic and static perspectives on field

following passage from a year seven science teacher comes towards the end of a number of lessons introducing the key components underpinning the seasons. Here, the teacher is speaking over an animation of the earth orbiting the sun. The important activities for explaining the seasons are in italics:

> Now when *the sun's rays hit the earth*, the earth has that also other theoretical midline, the equator, that breaks it into half. Northern hemisphere and southern hemisphere.
>
> ...
>
> Ok here we go. *The earth is spinning on its axis, happening every day.* Every time *it spins, it's moving day night, day night*. . . . And then *the earth is also moving around the sun.* You will notice at this point the northern hemisphere is closest to the sun, but when we started the southern hemisphere was closest to the sun. This is why we end up with opposite seasons. Because of that tilt in our axis puts us in different positions relationship to the sun.

The activities relevant here are to do with the Earth's rotation, both unmomented:

the earth is spinning;
it spins

and momented:

it's moving day night, day night

The passage also deals with the earth's orbit around the sun:

the earth is also moving around the sun

And it notes the emission of light from the sun to the earth:

the sun's rays hit the earth

These activities are crucial components for eventually showing that there are different temperatures at different times of the year and thus that there are seasons. This however does not capture the additional fact that different parts of the earth are affected differently. For this, the teacher additionally partitions the earth into a compositional taxonomy of northern and southern hemispheres, divided by the equator:

the earth has that also other theoretical midline, the equator, that breaks it into half. Northern hemisphere and southern hemisphere.

The activities of *orbit*, *rotation* and *light emission*, and the decomposition of earth into *northern* and *southern hemispheres* provide most of the information needed to explain the seasons. But the final sentence mentions one other crucial component not covered by the activity and item options in Figure 5.7, namely the *tilt* of the earth's axis. This tilt is crucial for any thorough explanation, as it accounts for the fact that different amounts of light hit different parts of the earth throughout the year. Indeed the following explanation by a student, which emphasizes tilt, is very positively evaluated by the teacher:

STUDENT   The seasons are created by the earth's 23.5 degree <u>tilt</u>. When the northern hemisphere is <u>tilted</u> towards the sun it is summer. As the earth orbits the sun the <u>tilt</u> stays the same, the side that's <u>tilted</u> towards the sun changes, making it winter in the northern hemisphere because it's furthest away.

TEACHER   Fantastic, I like that. That's a good one. All right. Hopefully yours says something similar to that.

Note that tilt is not an activity – it is not unfolding in any way and cannot be momented. Nor is tilt a part of the earth (we cannot say, for example, *the earth's tilt is a part of the earth*); nor is it a type of the earth. In order to account for tilt, we need to expand our model of field to introduce *properties*. To do this, we will take a step away from seasons into other areas of science and then return to how this affects the explanation of seasons considered above.

Property

In addition to activities and items, fields may be construed in terms of properties. Properties, in broad terms, organize potentially gradable qualities or positions that enable rich descriptions of phenomena. They often underpin distinctions between items and activities and are vital components of fields in themselves. They may characterize items, such as *the earth is _tilted_* or *the _negatively charged_ particle*; or they may qualify activities as in *the electrons oscillate _rapidly_*, or *inflation is _high_*.

If characterizing items, they can provide the criteria organizing taxonomies. For example, in nuclear physics, neutrons and protons are different types of nucleon that are distinguished primarily by their charge: protons are *positively charged*, while neutrons are *neutrally charged*. Similarly, though different bands of electromagnetic radiation such as visible light, ultraviolet light, x-rays and radio waves are all oscillating light, it is their wavelength that distinguishes them: visible light waves are around *390-700 nanometres long*, while ultraviolet waves are between *10-390 nanometres long*. In a similar fashion, properties may be key features distinguishing moments in an activity, for example, the activity of water freezing involves it getting *colder*, which means its particles move *more slowly*, and eventually they lock together to become ice.[7]

As properties can optionally occur for both activities and items, we can set up a basic network as in Figure 5.8.

Properties can take many forms. They may involve qualitative descriptions (*Everest is _tall_, it is climbed _slowly_*); or they may offer some spatio-temporal position (*Everest is _in Asia_, Everest's _current_ height, they trudged _through the snow_*). In addition, these properties may be graded and potentially ordered into *arrays* in relation to other properties (*Everest is _the tallest_ mountain on earth, The Himalayas stretch _from China to Pakistan through Bhutan, Nepal and India_*); this in turn opens the way for properties to be measured or quantified, what we will call *gauged* (*Everest is _8,848 m tall_, Everest is _27.99° N, 86.93° E_*).

Adding to our network, Figure 5.9 indicates that properties may be arrayed or not, and if arrayed, they may be gauged.

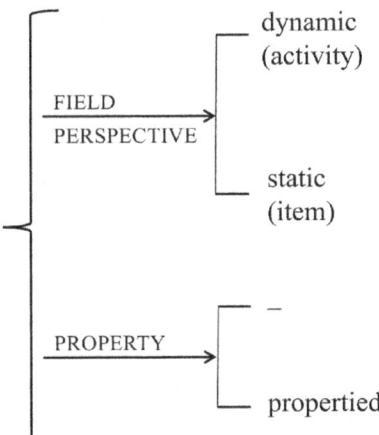

FIGURE 5.8 Basic parameters of field

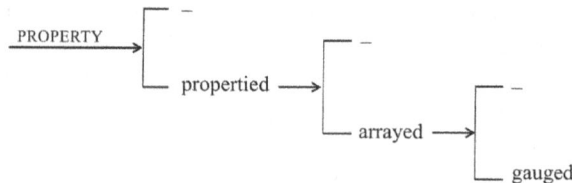

FIGURE 5.9   Network of PROPERTY, ARRAYING and GAUGING

Qualitative properties

We will explore properties through the following descriptive report from a field guide to Australian birds (Menkhorst *et al.* 2017: 472). In this text, properties permeate the description in order to evocatively describe the flame robin's physical qualities including its colour, shape and size. The properties not only allow birders to recognize flame robins, but also to distinguish between males and females, adults and juveniles, and other robins such as the dusky robin (bold and italics in original).

> **Flame Robin** *Petroica phoenicea*
> **Wing** 73-83 mm **Bill** 13-15 mm **Wt** 11-15g
> Largest, most slender-looking *Petroica* robin; long-bodied appearance accentuated by smallish head and longish wings. **Ad** ♂ distinctive; note *slaty upperparts; small white forehead spot; underparts bright orange-red from chin to belly.* **Ad** ♀ differ from Scarlet ♀ by *sandier brown* upperparts, *small* white forehead spot (sometimes tinged buff or wholly absent) and *light grey-brown breast* grading to whitish chin and belly. Some have diagnostic, but inconspicuous, traces of orange-yellow to orange on breast and belly. White eye-ring and bold wing-bars rule out Dusky Robin in Tas. **Juv** like Juv Scarlet with slightly finer white streaks on upperparts.
>
> **Voice:**   Contact call, a single note *tlip*, is sweeter than Scarlett Robin and seldom given on the non-breeding grounds. Musical warbling song, more complex and piping than other *Petroica* robins, often consists of 3 sets of 3 notes '*you may come, if you wish, to the sea*'.
>
> **Notes:**   Breeds mainly in upland eucalypt forests and woodlands, especially with open understory or small clearings; readily, but temporarily, colonizes cleared or burnt areas. Most leave the high country in autumn, wintering in more open habitats in lowlands including grasslands, farmland, and open forests and woodlands with grassy cover. Mostly seen singly or in pairs during the breeding season. Often seen in loose groups of up to 20 birds at other times, the only *Petroica* to form flocks.

We will focus initially on the first paragraph. This paragraph is primarily concerned with the physical appearance of the flame robin. As such it describes its properties such as colour, shape and size, and those of its various body parts (qualitative properties underlined):

<u>Largest</u>, <u>most slender-looking</u> *Petroica robin*
<u>long-bodied</u> *appearance*
<u>smallish</u> *head*
<u>longish</u> *wings*
<u>slaty</u> *upperparts*
<u>small</u> <u>white</u> *forehead spot*
underparts <u>bright orange-red</u> *from chin to belly*
<u>sandier brown</u> *upperparts*
sometimes <u>tinged buff</u> *or* <u>wholly absent</u>
<u>light grey-brown</u> *breast*
<u>whitish</u> *chin and belly*
traces of <u>orange-yellow</u> *to* <u>orange</u> *on breast and belly*
<u>White</u> *eye-ring*
<u>bold</u> *wing-bars*
<u>slightly finer</u> <u>white</u> *streaks on upperparts*

These precise verbal descriptions of colour (e.g. *slaty, whitish, sandier brown*), size (*largest, long-bodied, longish*) and shape (*most slender-looking*) are often supplemented by pictorial representations such as in Figure 5.10, to make it easier to recognize the bird in the wild (the original image is in colour, showing among other things, the distinct orange breast and belly of the adult male).

As this text shows, properties like colour, size and shape can be graded through *arraying*. In terms of field, arraying like this establishes degrees of a property in comparison to other more or less explicitly specified instances. For example, the size and shape of the flame robin is arrayed as:

<u>Largest</u>, <u>most</u> *slender-looking Petroica robin*

FIGURE 5.10 Picture of the flame robin showing its various properties. Reproduced from Menkhorst *et al.* (2017: 473) with permission from CSIRO Publishing. Original colour illustrations by Peter Marsack

Similarly, many of the colours are graded in terms of their brightness, or other qualities:

> underparts <u>bright</u> orange-red from chin to belly
> sand<u>ier</u> brown
> <u>light</u> grey-brown breast

If more precision is required, arrays can be gauged. The sub-heading shows this by gauging the length of the flame robin's wing and bill in terms of millimetres (mm) and its weight in terms of grams:

Wing <u>73-83 mm</u> **Bill** <u>13-15 mm</u> **Wt** <u>11-15g</u>

As students move through schooling, precise measurements of phenomena become more strongly emphasized. This is particularly the case for sciences such as physics where these measurements are regularly presented in graphs or long mathematical texts called quantifications (Lemke 1998; Parodi 2012; Doran 2018, Chapter 7 of this volume). The graph in Figure 5.11 is from a senior high-school physics exam. This graph arrays two properties of a wire on the vertical and horizontal axes, *resistance* and *temperature*, and gauges them in ohms (Ω) and degrees Celsius (°C) (*resistance* and *temperature* are in fact *itemized* properties, discussed below). A series of points (marked by X) and a trend line have been plotted by a student, giving specific measurements for both its resistance and temperature.

In a sense, properties, which may be arrayed and gauged, establish an ideational perspective on gradable meanings that have generally been explored through the interpersonal system of graduation within appraisal (Hood and Martin 2005, Martin and White 2005).[8] By putting forward the property network, we are suggesting that

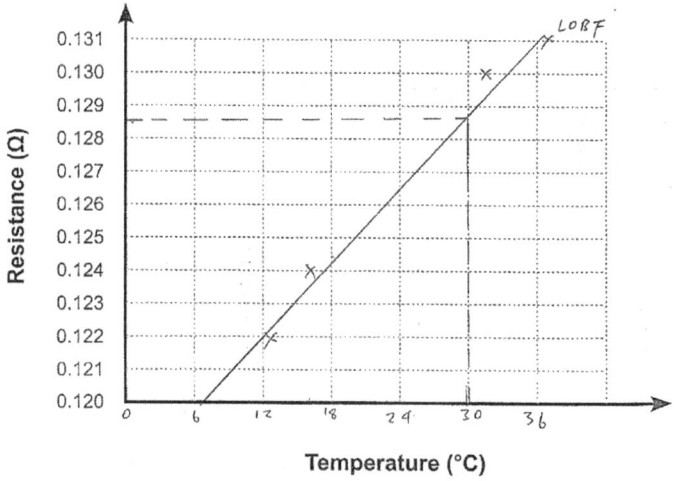

FIGURE 5.11 Graph of resistance and temperature, showing two gauged properties

the instances of grading shown above are not necessarily geared towards making evaluative meanings but are simply organizing the field of the *petroica robin*. It is likely that there will be some indeterminacy in this area, since properties supply an arena for attitudinal meanings invoked through graduation; indeed if we look at the second paragraph of the flame robin descriptive report, we see a handful of properties that could be read as invoked appreciations of the flame robin's call (graduated attitude, either inscribed or invoked, combined with arrayed properties underlined):

> **Voice:** Contact call, a single note *tlip*, is <u>sweeter than</u> Scarlett Robin and seldom given on the non-breeding grounds. Musical warbling song, <u>more complex and piping than</u> other *Petroica* robins, often consists of 3 sets of 3 notes '*you may come, if you wish, to the sea*'.

Spatio-temporal properties

A second type of property is concerned not with some quality of an item or activity, but rather with an item or activity's location in either space or time – termed *spatio-temporal* properties.

In our flame robin text, spatio-temporal properties are used extensively when detailing the flame robin's habitat and the time of year they can be observed (underlined):

> **Notes:** Breeds mainly <u>in upland eucalypt forests and woodlands</u>, especially with open understory or small clearings; readily, but temporarily, colonizes cleared or burnt areas. Most leave the high country <u>in autumn</u>, wintering <u>in more open habitats in lowlands including grasslands, farmland, and open forests and woodlands with grassy cover</u>. Mostly seen singly or in pairs <u>during the breeding season</u>. Often seen in loose groups of up to 20 birds <u>at other times</u>, the only *Petroica* to form flocks.

Like qualitative properties, spatio-temporal properties can occur for both activities and items. Examples of spatio-temporal properties of activities from the above excerpt include:

> *Breeds mainly <u>in upland eucalypt forests and woodlands</u>*
> *Most leave the high country <u>in autumn</u>*
> *wintering <u>in more open habitats</u>*
> *Mostly seen singly or in pairs <u>during the breeding season</u>*
> *Often seen in loose groups of up to 20 birds <u>at other times</u>*

Examples of spatio-temporal properties of an item from earlier in the text, include:

> *slightly finer white streaks <u>on upperparts</u>*[9]
> *Some have diagnostic, but inconspicuous, traces of orange-yellow to orange <u>on breast and belly</u>*

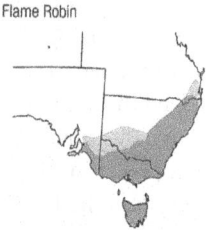

Flame Robin

FIGURE 5.12 Map showing spatio-temporal property of the flame robin in Australia. Reproduced from Menkhorst *et al.* (2017:472) with permission from CSIRO Publishing

Just like qualitative properties such as colour, size and shape, spatio-temporal properties can be arrayed by ordering their spatial or temporal positions in some way. In the initial description the underparts were described as:

> bright orange-red *from chin to belly*.

Here, the chin and belly are arrayed as outerpoints that the bright-orange red colour occurs between. Such arraying of spatio-temporal properties is often more easily shown through diagrams and maps, such as in Figure 5.12 showing the areas of Australia the flame robin can be observed.

Finally, like all properties, spatio-temporal properties can be gauged. An example of this is the use of latitude and longitude, such as 34.0386° S, 151.1407° E. Or more commonly in everyday discourse, this is regularly seen when giving the time:

> Be there *9 o'clock*.

From the description so far, we can flesh out our network of properties as Figure 5.13.

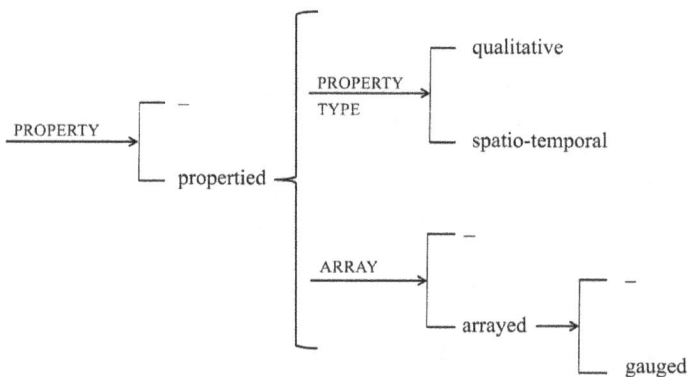

FIGURE 5.13 Network of PROPERTY

Complex fields

Activity, taxonomy and property systems provide the basic resources for construing field. But before we can return to mapping the explanation of seasons this chapter began with, there is one more system we need to introduce. This system makes room for interdependent variables that organize the complex constellations of meaning in technical discourses such as those of science. There are two basic ways of building interdependency: *reconstruing* and *interrelating*.

Reconstruing variables

Beginning with the reconstrual of field variables, we can return to the *tilt* of the earth that we mentioned is crucial for explaining seasons. In one sense, we can now readily account for this in our expanded model – *tilt* is a property of the earth and so can be arrayed and gauged. Indeed, our original text explaining the seasons does just that:

Seasons happen because Earth's axis is <u>tilted</u> at an angle of about <u>23.4 degrees</u>

In this form, where the *Earth's axis* is specified as being *tilted*, this analysis is unproblematic. However, in the next sentence, the property realized verbally as *tilted* above is nominalized as *tilt*:

Earth's axial tilt

In addition, the nominalized *tilt* is classified by *axial*. This positions tilt in a classification taxonomy; axial tilt is a type of tilt. This is problematic for the outline of field resources canvassed thus far as classification is a key feature of items, not properties. Moreover, in terms of Hao's model of ideational discourse semantics (2020a, Chapter 6 of this volume), the nominalized *axial tilt* is not a quality, which is the typical realization of property. Rather, it is a measured dimension of the earth that can be reorganized lexicogrammatically as a Focus^Thing structure:

Axial tilt of Earth

To reconcile these two seemingly conflicting analyses – *tilt* as a property and *tilt* as an item – what we will suggest here is that the property *tilted* is being reconstrued as an item *tilt*. This we will refer to as an *itemized property*. This analysis is based on the fact that *axial tilt* shows many of features of both properties and items: it can be arrayed and gauged like properties and can be taxonomized like items. And from the perspective of discourse semantics, it is not realized by a quality, which is the prototypical realization of qualitative properties, nor by an entity, which is the standard realization of items, but rather by a dimension of an entity.

Such itemized reconstruals are regularly used to 'name' broader sets of properties. Indeed, we have used a number of these throughout this paper to group together various properties. In the following examples, the underlined itemized properties are 'names' of the bolded properties:

The <u>colours</u> (of the flame robin) are **slaty, whitish** and **sandier brown**.
The <u>shape</u> (of the flame robin) is described as **slender-looking.**

Itemized properties comparable to these are the basis for symbolic variables in mathematics (see Doran, Chapter 7 of this volume). For example, in the following formula the *acceleration* of a moving body is symbolized as *a*, the *mass* of that body is *m* and the *force* on the body is *F*:

$$F = ma$$

Like all properties, each of these can be numerically gauged – in this case through Newtons (N), metres per second squared (m/s²) and kilograms (kg):

$$F = 4\,\mathrm{N}$$
$$a = 2\,\mathrm{m}\,/\,\mathrm{s}^2$$
$$m = 2\,\mathrm{kg}$$

But like items, they can all be taxonomized:

angular acceleration, linear acceleration;
centripetal force, electrostatic force;
relativistic mass, rest mass.

Reconstruals of properties as items is a regular feature of many fields. Under this interpretation, field is a resource not just for construing items, activities and properties, it is also a resource for reconstruing meanings. It enables multiple overlapping perspectives on phenomena to be realized in a single instance. This is emphasized by the fact that in addition to being itemized, properties can also be reconstrued as activities. Similarly, items can be dynamized by being reconstrued as activities, and activities can in turn be itemized. Each of these options have typical realizations in discourse semantics, as introduced below.

When properties are reconstrued as activities, such *activated properties* enable a dynamic unfolding of a property:

It *gets hotter*

In terms of property, this indicates an array of temperature (degrees of heat). But in terms of activity, it can be used to moment a larger activity:

It *gets hotter*
^
And then eventually it melts.

This type of activation moves us into the realm of 'becoming', where properties are, in effect, dynamized.

Activating items very commonly involves them being positioned in a taxonomy. The following excerpt from a university physics textbook shows an example of this for composition (Young and Freedman 2012: 742):

… now we let the ball touch the inner wall… The surface of the ball becomes part of the
cavity surface.

Here, the part whole relation between the surface of a ball and the cavity surface of
a wall is activated, and becomes a moment in a larger activity:

We let the ball touch the inner wall
∧

The surface of the ball becomes part of the cavity surface.

Finally, just as items can be activated, activities can be itemized.[10] This regularly
happens in the process of technicalization, as scientific terms are distilled as activity
entities (see Hao 2020a and Chapter 6 of this volume for discussion of the discourse
semantics of activity entities). One of our initial examples of a momented activity in
fact showed a series of itemized activities that moment *phagocytosis* – itself an itemized activity; the itemized activities involved are underlined below:

Detection
∧

Ingestion
∧

Phagosome forms
∧

Fusion with lysosome
∧

Digestion
∧

Discharge

Like all items, these itemized activities can enter into a taxonomy. For example,
phagocytosis is one type of *endocytosis*, along with *potocytosis* and *micropinocytosis*, which
all contrast with *exocytosis*.

Table 5.1 shows various reconstruals and some typical discourse semantic realizations (from Hao 2020a). Note here that the order in which the reconstrual takes
place is significant – an itemized activity is different from an activated item.

TABLE 5.1 Field reconstruals

Field reconstruals	Typical discourse semantic realization	Example
itemized property	measured or perceived dimension	The *colour* of skin
activated property	occurrence figure	It heats up
	state figure	It gets hotter
itemized activity	activity entity	Phagocytosis
activated item	state figure	You become part of the team

Interrelating fields and the explanation of seasons

We are now in a position to characterize each of the main elements of our initial explanation of the seasons, replayed here:

What Causes Seasons on Earth?

Seasons happen because Earth's axis is tilted at an angle of about 23.4 degrees and different parts of Earth receive more solar energy than others.

Because of Earth's axial tilt (obliquity), our planet orbits the Sun on a slant which means different areas of Earth point toward or away from the Sun at different times of the year.

Around the June solstice, the North Pole is tilted toward the Sun and the Northern Hemisphere gets more of the Sun's direct rays. This is why June, July and August are summer months in the Northern Hemisphere.

Opposite Seasons

At the same time, the Southern Hemisphere points away from the Sun, creating winter during the months of June, July and August. Summer in the Southern Hemisphere is in December, January and February, when the South Pole is tilted toward the Sun and the Northern Hemisphere is tilted away.

For our discussion here, the main field components needed for this explanation are:

> *Earth's 23.4 degree axial tilt*
> *Division of earth into northern and southern hemispheres*
> *Earth's orbit of the sun*
> *Earth receipt of solar energy*
> *Different parts of the earth receive more solar energy than others at different times of the year*
> *Seasons of summer and winter*

Each of these components is classified with reference to field resources in Table 5.2. For ease of reference, each of the components is named in small caps (TILT, HEMISPHERES, etc.).

The first thing to note is that far more than a common sense understanding of the seasons is involved. For a full explanation, students need to attend to both the classification and composition of items, two types of activity and both gauged and arrayed properties. In addition, they need to conceptualize three itemized properties, and a set of items ordered into two separate arrays. This is already a conceptual challenge for most junior secondary school students.

But there is more going on. As set out in Table 5.2, more than one factor is needed for the explanation, however we have not yet specified the relations among

TABLE 5.2 Field relations in an explanation of the seasons

Factors in explanation of seasons	Field relation
Tilt	
Earth's 23.4 degree axial tilt	classified itemized gauged property
Hemispheres	
Division of earth into northern and southern hemispheres	composition taxonomy of items
Orbit	
Earth's orbit of the sun	itemized cyclical activity
Receipt of solar energy	
Earth receives solar energy	culminative activity
Variation in solar energy	
Different parts of the earth receive more solar energy than others at different times of the year	composition\ taxonomy (*different parts of the earth*) ordered into an array of an itemized qualitative property (*more solar energy*) ordered into another array of an itemized spatio-temporal property (*different times of the year*)
Seasons	
Seasons of summer and winter	classification taxonomy of items

them. In order to explain seasons, specific types of logical interdependency have to be established. To model this, we will introduce the second way of linking field variables – *interrelating* (complementing reconstruing as introduced above).

Interrelating is concerned with how different elements of field are associated with each other. In broad terms, there are two means of interrelating elements: they can be positioned as relatively independent of one another or they can be positioned as in some sense dependent on each other (although not precisely the same, this closely relates to the LCT distinction between dependent and independent links in constellations, introduced in Maton and Doran, Chapter 3 of this volume).

Let's focus first on the relatively independent relation. In the seasons explanation, the TILT, HEMISPHERES, ORBIT and RECEIPT OF SOLAR ENERGY are not dependent on each other in any way. One can vary or not exist at all without affecting the others. The fact that the earth is tilted, for example, has no bearing on the fact that the sun emits solar energy that hits the earth. Similarly, the fact that there are hemispheres on earth has no impact on whether it orbits the sun.

To describe this relation, we can borrow from Halliday and Matthiessen's (2014) logico-semantic relations for English clause complexing and call this *extension* (signified by a +). Here we are analogizing from an 'and' relation, where multiple elements are coordinated but are not ordered in any way (ideationally speaking). The extending factors in this relationship can be usefully laid out in parallel, as follows:

TILT	+	HEMISPHERES	+	ORBIT	+	RECEIPT OF SOLAR ENERGY
itemized property		composition of items		cyclical activity		culminative activity

In contrast, some elements of field are dependent on others. For example, the VARIATION IN SOLAR ENERGY factor is the result of the combination of the TILT, ORBIT, HEMISPHERES and RECEIPT OF SOLAR ENERGY factors. Without each of these factors, the precise form of the VARIATION IN SOLAR ENERGY would not occur. This in part accounts for the complexity of seasons explanations. VARIATION IN SOLAR ENERGY takes the compositional taxonomy from the HEMISPHERES, orders it into an array of *solar energy* involved in the RECEIPT OF SOLAR ENERGY, reinterprets the ORBIT of the sun as an array of time through the year and, through a couple of unspecified steps resulting from the TILT, additionally arrays the variation in solar energy at different parts of the earth according to the time of the year. Analogizing again from Halliday and Matthiessen's logico-semantic relations, we will call this dependency relation *enhancing* (signified by an x).

Finally, the SEASONS reorganizes the VARIATION IN SOLAR ENERGY from an array of temperatures to a classification taxonomy of items – summer, winter, autumn, spring. This in effect names components of these variations and distils its meaning into the technical term 'seasons'. Following our analogy with Halliday and Matthiessen's logico-semantic relations, we will call this naming relation *elaboration* (signified by =).

Pulling these interrelation types together with our previous description, we can visualize the field variables underpinning the explanation of the seasons as follows (with enhancing factors laid out vertically in relation to what they depend upon):

TILT	+	HEMISPHERES	+	ORBIT	+	RECEIPT OF SOLAR ENERGY
itemized property		composition of items		cyclical activity		culminative activity
				×		

VARIATION IN SOLAR ENERGY

compositional taxonomy on an arrayed itemized property on another arrayed itemized property

=

SEASONS

classification taxonomy of items

This outlines how the relatively independent components of the TILT, HEMISPHERES, ORBIT and RECEIPT OF SOLAR ENERGY together produce the VARIATION IN SOLAR ENERGY, which in turn produces the SEASONS. The mapping effectively displays the complexity underpinning a scientific explanation of a phenomenon we all experience in everyday terms.

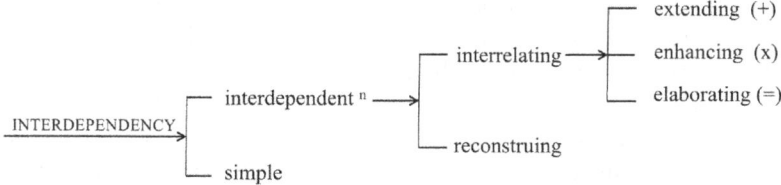

FIGURE 5.14 Network for INTERDEPENDENCY

As far as our broader model of field is concerned, this completes the description. The options for building interdependency in fields produces the network in Figure 5.14.

Pulling this together with the rest of the field network gives the basic systems of field shown in Figure 5.15. This network indicates that construals of field can adopt either a dynamic or a static perspective, can be optionally propertied and can be related to other aspects of a field.

Making uncommon sense

From the perspective of educational linguistics, our model of field as a resource for construing phenomena reveals the complexity of science fields at even the lower levels of secondary school. The explanation that we focused on for this paper is a relatively simple one in the broader scheme of things. It did not take into account

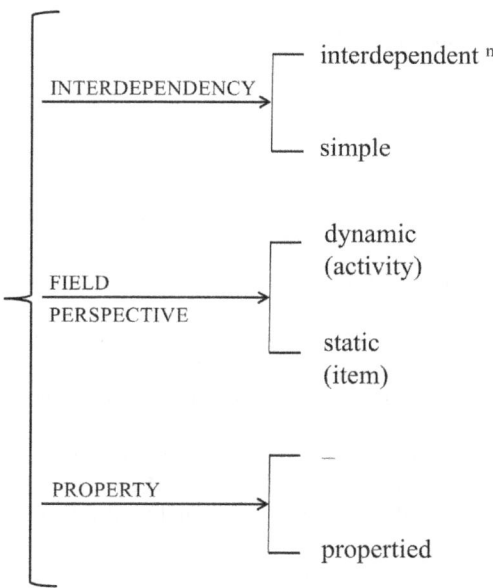

FIGURE 5.15 Network of field

the effect of variation in the length of the day arising from the rotation of the earth; nor did it go into detail about how the earth's tilt actually produces variation in light intensity in different part of the earth at different times of the year; nor did it conceptualize the hemispheres as a continuous array of latitude (affecting how 'summery' or 'wintery' it is depending on how far you are from the poles); nor did it consider this explanation in relation to other meteorological and astronomical phenomena. And it definitely did not consider how this conception of seasons is related to seasonal change in parts of the world where a wet-dry distinction is more relevant (and in doing so can be critiqued from a post-colonial perspective). As scientific knowledge develops through schooling, more and more of these additional elements are interrelated and then presumed. For an educational program that aims to develop a discipline-specific pedagogy based on the varying ways knowledge is built, managing this complexity such as through a model of this kind is crucial.

As far as functional linguistics is concerned, our model of field builds on Hao's modelling of ideational discourse semantics (2020a) and opens up the possibility of explicitly differentiating and linking ideational meaning resources across strata (register, discourse semantics and lexicogrammar – specifically, field realized through ideation and connexion, and ideation and connexion realized through clause complexing, transitivity and nominal group structure). It has done so in a way that is proving productive not just for language, but for a range of semiotic resources used in science (see Doran, Chapter 7 of this volume). This places functional linguistics and semiotics in a far stronger position to manage the distinctive complexity of knowledge building across disciplines, and within disciplines across modalities of communication. Making sense of uncommon sense depends on robust modelling of this kind.

Notes

1 https://www.timeanddate.com/astronomy/seasons-causes.html
2 A little closer to everyday life, following Barthes (1977: 101), we could alternatively moment the activity *having a drink,* into the series *ordering a drink, obtaining it, paying for it, drinking it.*
3 Previously, series of activities were termed *activity sequences.* However in light of Hao's work on ideational discourse semantics (2015, 2018, 2020a), *series* will be used here for strings of activities in field and *sequences* will be reserved for strings of figures in discourse semantics.
4 This and a number of other examples in this chapter come from a study examining classroom practice in secondary schools in New South Wales, Australia, led by Karl Maton, J. R. Martin, Len Unsworth and Sarah Howard and funded by the Australian Research Council (DP130100481).
5 We will again follow Hao (2018, 2020a) here in using CONNEXION for the discourse semantic relation between figures, previously known as CONJUNCTION in Martin (1992) and Martin and Rose (2007). This is in order to distinguish the discourse semantic system from the word-class conjunction within lexicogrammar.

6 Here *tides* is an activity entity that is caused by the previous occurrence figures. However, under Hao's model (2020a: 94), it is in fact part of a presented state figure within the larger sequence, alternatively realised as *The gravitational pull of the sun and the moon and the rotation of the earth causes (the forming of) tides* (Hao personal communication). Hao argues that sequences such as this are one means through which activity entities like *tides* can be presented in scientific texts.

7 Thanks to Dragana Stosic, Sally Humphrey and Jing Hao for this example. This raises the broader point that all perspectives on field (activity, taxonomy, property) can be used as criteria for distinguishing and defining all other perspectives. For any particular field, its interlocked networks of field relations are such that its technical meanings will be mutually defining.

8 A complementary ideational perspective on gradable meanings from the perspective of discourse semantics is given by Hao (2015, 2020a) and Hood and Hao (chapter 10, this volume), in terms of qualities of entities.

9 This is not saying that the upperparts are white (i.e. it is not attributing a qualitative property to the upperparts). Rather is locating the white streaks as being *on* the upperparts.

10 Note that we do not need to account for *propertied items* or *propertied activities* in this section as this is already accounted for in our initial system in Figure 5.8, where either an activity or an item can take a property.

References

Barthes, R. (1977) *Image, Music, Text*, London: Fontana.

Doran, Y. J. (2017) 'The role of mathematics in physics: Building knowledge and describing the empirical world', *Onomázein Special Issue*, March: 209–26.

Doran, Y. J. (2018) *The Discourse of Physics: Building Knowledge through Language, Mathematics and Image*, London: Routledge.

Doran, Y. J. (2019) 'Building knowledge through image in physics', *Visual Communication*, 18(2): 251–77.

Doran, Y. J. (2020) 'Cultivating values: Knower-building in the humanities', *Estudios de Lingüistica Aplicada*, 37(70): 169–98..

Greenwood, T. and Allen, R. (2004) *Year 12 Biology 2004: Student resource and activity manual*, Hamilton: Biozone.

Halliday, M. A. K. (2004) *The Language of Science: Volume 4 in the Collected Works of M. A. K. Halliday*, London: Continuum.

Halliday, M. A. K. and Martin, J. R. (1993) *Writing Science: Literacy and Discursive Power*, London: Falmer.

Halliday, M. A. K. and Matthiessen, C. M. I. M. (2014) *Halliday's Introduction to Functional Grammar*, London: Routledge.

Hao, J. (2015) '*Construing Biology: An Ideational Perspective*', Unpublished PhD thesis, Department of Linguistics, University of Sydney.

Hao, J. (2018) Reconsidering "cause inside the clause" in scientific discourse – from a discourse semantic perspective in systemic functional linguistics. *Text and Talk*, 38(5): 520–50.

Hao, J. (2020a) *Analysing Scientific Discourse from a Systemic Functional Perspective: A Framework for Exploring Knowledge Building in Biology*, London: Routledge.

Hao, J. (2020b) 'Construing relations between scientific activities through Mandarin Chinese', in J. R. Martin, Y. J. Doran and G. Figueredo (eds) *Systemic Functional Language Description*, London: Routledge, 238–72.

Hao, J. (2020c) 'Nominalisation in scientific English: a tristratal perspective', *Functions of Language*, 27(2): 143–73.

He, Y. (2020) *'Animation as a Semiotic Mode: Construing Knowledge in Science Animated Videos'*, unpublished PhD thesis, University of Sydney.

Hood, S. (2010) *Appraising Research: Evaluation in Academic Writing*, New York: Palgrave Macmillan.

Hood, S. and Martin, J. R. (2005) 'Invoking attitude: The play of graduation in appraising discourse', *Revista Signos*, 38(58): 195–220.

Lemke, J. L. (1990) *Talking Science: Language, Learning and Values*, Norwood, NJ: Ablex.

Lemke, J. L. (1998) 'Multiplying meaning: Visual and verbal semiotics in scientific text', in J. R. Martin and R. Veel (eds) *Reading Science*, London: Routledge, 87–113.

Martin, J. R. (1992) *English Text: System and Structure*, Amsterdam: John Benjamins.

Martin, J. R. (2011) 'Bridging troubled waters: Interdisciplinarity and what makes it stick', in F. Christie and K. Maton (eds) *Disciplinarity*, London: Routledge, 35–61.

Martin, J. R. (2020) 'Revisiting field: Specialized knowledge in secondary school science and humanities discourse', in J. R. Martin, K. Maton and Y. J. Doran (eds) *Accessing Academic Discourse*, London: Routledge, 114–47.

Martin, J. R. and Maton, K. (2013) (eds) 'Cumulative knowledge-building in secondary schooling', Special Issue of *Linguistics and Education*, 24(1): 1–74.

Martin, J. R. and Rose, D. (2007) *Working with Discourse: Meaning beyond the Clause*, London: Continuum.

Martin, J. R. and Rose, D. (2008) *Genre Relations: Mapping Culture*, London: Equinox.

Martin, J. R. and Veel, R. (eds) (1998) *Reading Science: Critical and Functional Perspectives on the Discourse of Science*, London: Routledge.

Martin, J. R. and White, P. R. R. (2005) *The Language of Evaluation: Appraisal in English*, Basingstoke: Palgrave Macmillan.

Martin, J. R., Maton, K. and Doran, Y. J. (2020) *Accessing Academic Discourse: Systemic functional linguistics and Legitimation Code Theory*, London: Routledge.

Martin, J. R., Maton, K. and Matruglio, E. (2010) 'Historical cosmologies: Epistemology and axiology in Australian secondary school history discourse', *Revista Signos*, 43(74): 433–63.

Martin, J. R., Maton, K. and Quiroz, B. (eds) (2017) 'Systemic functional linguistics and Legitimation Code Theory on education and knowledge', *Onomázein Special Issue*, March: 1–242.

Martin, J. R., Unsworth, L. and Rose, D. (in press) 'Condensing meaning: Imagic aggregations in secondary school science', in G. Parodi (ed) *Multimodality*, London: Bloomsbury.

Maton, K. (2014) *Knowledge and Knowers: Towards a Realist Sociology of Education*, London: Routledge.

Maton, K. and Doran, Y. J. (2017) 'SFL and code theory', in T. Bartlett and G. O'Grady (eds) *The Routledge Handbook of Systemic Functional Linguistics*, London: Routledge, 605–18.

Maton, K., Martin, J. R. and Matruglio, E. (2016) 'LCT and systemic functional linguistics: Enacting complementary theories for explanatory power', in K. Maton, S. Hood and S. Shay (eds) *Knowledge-Building*, London: Routledge, 93–113.

Menkhorst, P., Rogers, D., Clarke, R., Davies, J., Marsack, P. and Franklin, K. (2017) *The Australian Bird Guide*, Clayton South: CSIRO Publishing.

Parodi, G. (2012) 'University genres and multisemiotic features: Accessing specialized knowledge through disciplinarity', *Fórum Linguístico*, 9(4) 259–82.

Unsworth, L. (2020) 'Intermodal relations, mass and presence in school science explanation genres', in M. Zappavigna and S. Dreyfus (eds) *Discourses of Hope and Reconciliation*, London: Bloomsbury, 131–52.

Wignell, P., Martin, J. R. and Eggins, S. (1989) The discourse of geography: Ordering and explaining the experiential world, *Linguistics and Education*, 1(4): 359–91.

Young, H. D. and Freedman, R. A. (2012) *Sear's and Zemansky's University Physics with Modern Physics*, 13th ed., San Francisco: Addison Wesley.

6

BUILDING TAXONOMIES

A discourse semantic model of entities and dimensions in biology

Jing Hao

Introduction

Studying science at the undergraduate level represents a critical stage of apprenticeship into scientific disciplines. Undergraduate students who are training in the biological sciences learn to observe biological phenomena, conduct experiments, engage with academic literature, record observations and provide reasoning largely through written texts. Through this work students make a transition from knowledge 'reproduction' to knowledge 'production' (Bernstein 2000), including a shift from building knowledge that is recontextualized in pedagogic texts to engaging with knowledge that is published in peer-reviewed research articles. In Australian universities, students who have been successfully apprenticed into undergraduate course work can choose a pathway towards postgraduate research. This chapter explores one aspect of knowledge development during this transition, that of building taxonomies.

As introduced in Doran, Maton and Martin (Chapter 1, this volume), understanding scientific knowledge building has been a longstanding object of study in both Systemic Functional Linguistics (SFL) and Legitimation Code Theory (LCT). SFL conceptualizes scientific taxonomies by and large from an ideational perspective, in terms of ideational meanings construing the register variable field. Scientific taxonomies are characterized as a key dimension of 'technicality' in science (Martin 1992; Halliday and Martin 1993). Technical taxonomies are distinguished from those used in other fields, such as everyday life and recreational activities. Their differences are determined largely by different recognition criteria in different fields. Technical taxonomies (e.g., cells, pathogen, bacteria) are identified based on criteria that are typically learned in the institutional settings, such as through scientific measurements. Taxonomies in the recreational fields are identified based on specialized usage and functionality (e.g., shuttlecock, screwdriver, piano); and in everyday life taxonomies are identified based on our daily and domestic experience (e.g., shoes, toothbrush, bags). The differences of fields recognized in SFL resonates with Bernstein's (2000) sociological distinction between 'horizontal discourse' and 'vertical discourse' (i.e. common sense vs. uncommon sense), and in vertical discourse between 'horizontal knowledge structures' (e.g., the humanities) and 'hierarchical knowledge structures' (e.g., sciences). Both SFL's field typology and Bernstein's concepts provide useful ways to think about the nature of different disciplines. However, when it comes to analyzing knowledge structures in practice, neither of them provide sufficient analytical tools. As Maton (2014) argues, Bernstein's concepts of discourses and knowledge structures are 'good to think with but less useful to analyze with' (Martin and Maton 2017: 28).

LCT extends Bernstein's concepts and offers a rigorous multi-dimensional conceptual toolkit for analyzing disciplinary knowledge building (Maton 2014, Maton *et al.* 2016). Building scientific taxonomies, for instance, can be analyzed through the theoretical dimensions of both Specialization and Semantics (Maton 2020, Maton and Chen 2020). With respect to Specialization, biological science can be seen as a *knowledge code*, emphasizing stronger epistemic relations between scientific claims and the biological phenomena under study, and sfvweaker social relations between scientific claims and biologists themselves. In terms of Semantics, *semantic density*, particularly its form of *epistemological condensation*, offers a way of conceptualizing the degree to which meanings are condensed in building taxonomies. Meanings at stake include the role of taxonomies in biological activities (Martin and Maton 2013).

This study is an attempt to make knowledge structure visible from the perspective of SFL. A key question is how to reveal disciplinary knowledge (i.e. field) by analyzing language patterns in texts. The language features concerned are particularly associated with analyses of unfolding meaning across texts – a **discourse semantics** perspective, which addresses meaning made in and beyond the clause. Descriptions of one discourse semantic unit, known as *entity*, have been used to identify the language of taxonomies in field (Martin and Rose 2007; Martin 1997). However, the classification of entities has by and large drawn on distinctions in the register variable field (Martin 1992). This limits the usefulness of discourse semantics as a tool for examining the breadth and depth of taxonomies.

To address both theoretical and descriptive challenges, this chapter reports on further development of an SFL framework for analyzing scientific taxonomies. I first introduce the theoretical principle known as 'trinocularity', whereby language choices are viewed from three simultaneous perspectives (Matthiessen and Halliday 2009; Martin 2013). I then draw on the trinocular principle to describe a range of choices in the system network of entities, including the systems of ENTITY TYPE and DIMENSIONALITY. Note that since the categories have emerged from undergraduate biology texts (including pedagogic materials and students' experimental reports) (Hao 2015), the description does not aim to provide generalized categories that can be used for all scientific discourse. However, the method through which the choices are identified *is* appliable and can be generalized across studies. After introducing the systems, I illustrate how the framework outlined can be used for analyzing texts. An excerpt from a high-graded student research report produced at the final undergraduate year is selected for the illustrative analysis. The analysis reveals both the diversity and depth of taxonomies that are developed at the end of undergraduate study.

Foundations

Theoretical principles of language description

A critical understanding of language from an SFL perspective is that language is a primary means for construing knowledge (together with other semiotic resources such as images and body language, see Hood and Hao this volume). 'Construe' means that knowledge is not only expressed by language (and other semiotics), but its organization is also influenced by how language is used (Halliday 2004 [1998]). A linguistic understanding of scientific taxonomies involves recognizing language resources that construe taxonomies.

SFL theorizes the language of disciplinary knowledge from a multi-stratal and multi-functional perspective. The theoretical dimensions of stratification and metafunction are modelled visually in Figure 6.1.

Stratification conceptualizes the inherent relationship between language and context, by recognizing patterns of meaning across different levels of abstraction known as strata. The strata within language include **phonological/graphological** systems organizing sound/script; **lexicogrammatical systems** organizing meanings in a clause; and **discourse semantic systems** organizing meaning unfolding in a text. The strata are simultaneously organized from the perspective of three metafunctions,

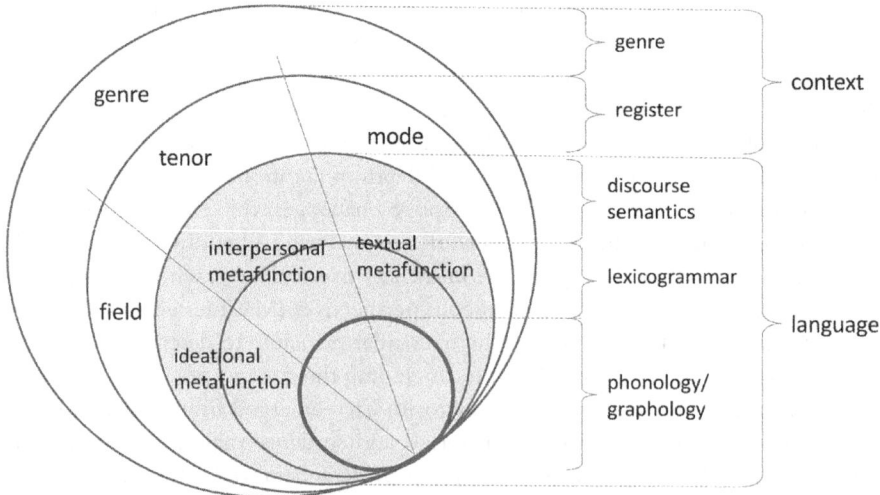

FIGURE 6.1 Stratification and metafunctions in SFL theory

which relate naturally to three contextual register variables – **ideational meanings** construing what is going on in a **field**; **interpersonal meanings** enacting social relations (**tenor**); and **textual meanings** composing **mode** (e.g., spoken vs. written discourse). Interactions among field, tenor and mode enable the performance of a range of staged and goal-oriented social purposes in our culture – i.e. **genres** (Martin and Rose 2008).

Critical to understanding both stratification and metafunction is the idea that choices in language are not made in isolation but are based on: 1) meaning that is construed at a higher stratum, 2) meaning at a lower stratum that is a realization of the meaning at a given stratum, and 3) meanings in the other metafunctions at a given stratum. This simultaneous consideration of meaning-making from 'above', 'below' and 'around' is in essence a **tri-stratal** perspective. Specific concerns are determined by specific descriptive tasks. If our standpoint is ideational discourse semantics, of particular concern are meanings construed at the level of field (looking from 'above'), their realizations through experiential grammar (looking from 'below'), and the interactions of ideational with interpersonal and textual meanings in the discourse (looking from 'around'). For further work adopting a tri-stratal perspective, see (Hao 2020a) on 'nominalization' and (Hao 2018) on 'causality'.

In what follows, I review previous work on scientific taxonomies. I argue that previous descriptions were by and large based on field and grammar – reasoning from above and below – but not from around. There is accordingly a need to bring discourse semantic resources into the picture.

Previous SFL description of scientific taxonomies

Taxonomies at the level of field

From the perspective of the register variable field, Martin (1992) models taxonomy as one aspect of field organization, along with activity sequences the taxonomies enter into. Distinctive configurations of activity sequences and taxonomies allow several field types to be identified, as shown in Table 6.1.

Science is seen as an example of a 'technical' field (Martin 2017), characterized by 'technical' taxonomies and 'implication sequences'. The term 'implication'

TABLE 6.1 Field types in relation to activity sequences and taxonomies (c.f. Martin 1992: 545, Martin 2017)

	activity sequences	*taxonomies*
domestic (guidance)	implicit	'natural'
specialized (participation)	manuals	utilitarian (tools)
e.g. sport, hobby, trades		
administration (cooperation)	procedures	pragmatic (subjects)
exploration / technical (instruction)	implication	technical (things)
e.g. humanities, social sciences and science	sequence	

emphasizes that in scientific explanations one activity is determined by what has gone before. The term 'technical' emphasizes that scientific taxonomies have distinctive criteria for classification (superordinate and subtypes) and composition (parts and whole) – different from those used in everyday and specialized fields. Wignell *et al.* (1993) exemplify this difference by comparing taxonomies of *birds of prey* in science with those recognized in the specialized field of bird watching. They show that birdwatchers' taxonomy relies largely on observable physical characteristics (e.g., colour and tail-shape for *Black Kite vs. Square-tailed Kite* in Figure 6.2), whereas scientific taxonomies for birds of prey considers their differences with respect to chromosomes and genes. Scientific taxonomies also often involve supplementing vernacular names with binomial Latin terminology (e.g., *Acciptridae*, *Milvus*, *Lophoictinia* in Figure 6.3).

Complementing Martin's (1992) field typology, Hood (2010) recognizes that in academic texts, specifically research articles, two fields can be identified based on their distinctive taxonomies and activity sequences: one is the field of the 'object of study' that is explored by the discipline (e.g. the scientific phenomena observed by scientists); the other is the field of 'research' referring to 'the process of enquiry and knowledge building' (e.g. the scientific methods practised by scientists) (2010: 121). This understanding helps us to make an important distinction between biology as a body of knowledge (i.e. the object of study) and biology as an academic discipline (involving both the object of study and research). This chapter adopts Hood's distinction, considering taxonomies for both the object of study and research.

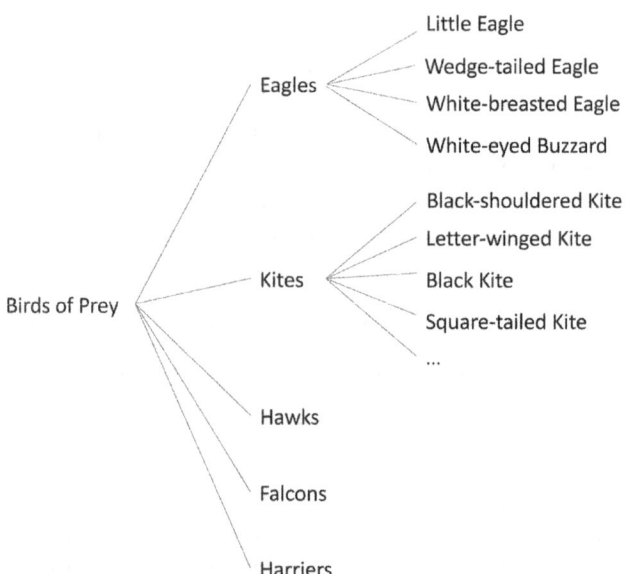

FIGURE 6.2 A simplified birdwatchers' taxonomy of *birds of prey* (adapted from Wignell *et al.* 1993: 156)

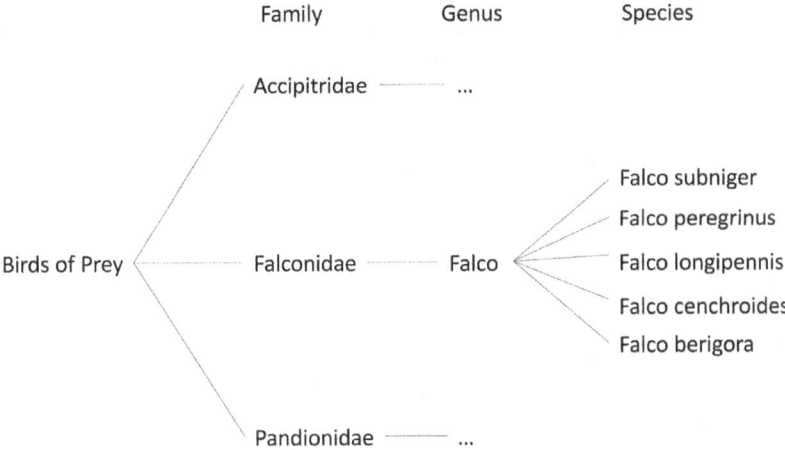

FIGURE 6.3 A simplified scientific taxonomy of *birds of prey* (adapted from Wignell *et al.* 1993: 157)

Nominal realizations of taxonomies

Alongside the field perspective discussed above, previous studies have also adopted a lexicogrammatical perspective on scientific taxonomies. It was recognized that taxonomies are realized nominally, typically through a Thing (e.g. *cell*) or a Classifier^Thing (e.g. *cell wall*) structure. Halliday suggests that this is because experiential meanings construed by nominal groups are 'more stable' and 'likely to be subcategorized' in comparison to other word classes (Hao 1998: 197). For this reason, realizing taxonomies nominally implicates the transcategorization potential of grammar – especially deriving nouns from other word classes (i.e. nominalization). Naming a person who *runs* as a *runner* or a person who *cooks* as a *cook* both involve transcategorizing a verb as a noun.

In addition to its vocabulary building function, transcategorization affords the possibility of remapping the realization of discourse semantics in lexicogrammar. For example, a discourse semantic **figure** (Halliday and Matthiessen 1999; Hao 2020b), which is typically realized through a clause (e.g., *the salt diffused in the water*), can be remapped as a nominal group (e.g., *the diffusion (of salt in the water)*). The remapping of a figure as a nominal group creates **stratal tension** between discourse semantics and lexicogrammar, known as **grammatical metaphor** – here, to be more specific, an experiential metaphor (Halliday 1994; Martin 2008).[1]

As scientific knowledge develops, what was initially an experiential metaphor can be gradually 'distilled' (Martin 1993: 191) to become a conventional way of talking about activities in a particular field; it thus loses stratal tension, becoming a 'dead metaphor' (Halliday 1998a). In science, distilled technical terms are typically introduced by definition. For example, *diffusion,* which we treated above as a grammatical metaphor, can be defined as in Table 6.2. A definition is structured through an elaborating relation of some kind – commonly via a Token^Value structure realizing a relational identifying clause.

TABLE 6.2 A linguistic definition of *diffusion*

Term		Definition
diffusion	*is*	*the process whereby a substance in high concentration moves to a place of low concentration*
Token	Process: intensive identifying	Value

The linguistically defined technical terms are often involved in construing tax-onomies of activities (e.g. Halliday 1998a; Wignell *et al.* 1993; Martin and Rose 2007). This understanding connects meanings at the level of field with grammar. We now turn to how resources in between field and grammar – i.e. discourse semantic systems – were considered in previous studies.

Previous discourse semantic descriptions of entities

At the level of discourse semantics, language for realizing taxonomies has been pre-viously discussed in relation to the discourse semantic system of IDEATION (Martin 1992). The description related to the realization of taxonomies is the classification of entity types presented in Martin and Rose (2007), reproduced below in Table 6.3.

While the entity types in Table 6.3 are intended to cover a diversity of realiza-tions, this categorization is problematic in many ways. First, most of the categories are motivated largely from the perspective of field. The more delicate entity types (i.e. everyday, specialized, technical and institutional) are in an one-to-one corre-spondence to the field types presented in Martin (1992) (i.e. domestic, specialized, administrative and exploration). Based on these categories, entities used in scientific discourse (e.g. *cells, gene, fungi, inflation*) are seen as 'technical', under the general cat-egory of 'abstract'. This creates a number of problems in the application of these cat-egories. For example in biological science, items are often tangible and observable, with concrete physical presence (e.g. *Birds of prey* exemplified in Wignell *et al.* 1993). Generalizing these as 'abstract' meanings is misleading with respect to the tangibility of observable scientific phenomena. In addition, the one-to-one mapping of field types and entity types creates uncertainty when determining whether the identified 'entities' represent discourse semantic features or field types.

Second, the category 'metaphoric' treats grammatical metaphors as 'entities', which contradicts the concept of grammatical metaphor as stratal tension between discourse semantics and lexicogrammar, rather than a meaning *on* a particular stratum. This creates serious confusion in relation to the discourse semantic meanings at stake.

Third, while not explicitly specified in Table 6.3, technical entities also include elements referring to activities (e.g., *inflation*). No criteria, however, are offered as to how the nominalizations which function as 'technical' entities are to be distin-guished from those that are experiential metaphors (e.g. *exposure* and *humiliation* in Table 6.3). And no clear distinction is made with respect to the technical entities referring to scientific items (e.g., *gene, cells*).

TABLE 6.3 Kinds of entities (Martin and Rose 2007: 114)

Indefinite pronouns		*some/any/nothing/one*
concrete	everyday	*man, girlfriend, face, hands, apple, house, hill*
	specialized	*mattock, lathe, gearbox*
abstract	technical	*inflation, metafunction, gene*
	institutional	*offence, hearing, applications, violation, amnesty*
	semiotic	*question, issue, letter, extract*
	generic	*colour, time, manner, way, kind, class, part, cause*
metaphoric	process	*relationship, marriage, exposure, humiliation*
	quality	*justice, truth, integrity, bitterness, security*

Furthermore, the instances of 'generic' entity are often realized through the Focus function in a nominal group structure, in form of an embedded nominal group (e.g. *kind* in *the kind of; part* in *the part of*). By identifying the meaning construed by Focus group as a distinctive entity type over-privileges the perspective from grammar and is inconsistent with Martin's (1992: 314) suggestion that a Focus^Thing structure as a whole enters into lexical cohesion.[2] So in his terms, a Focus^Thing structure realizes one discourse semantic unit, rather than two.

As a result of these inconsistencies, the framework shown in Table 6.3 is not readily applicable to the analysis of the discourse semantics of ideational meaning. A more appliable discourse semantic description is needed in order to understand how taxonomies are construed by language. In the following sections, I introduce a refined discourse semantic system of entities for analyzing taxonomies, including types of entities and their associated dimensions (cf. Hao 2015).

A description of discourse semantic entities

The discourse semantic description of entities presented here is based on a range of scientific texts used in undergraduate biology, including pedagogic materials such as textbooks and laboratory manuals, and high-graded students' written assessments across undergraduate years (including 26 laboratory reports and six research reports). I consider entities from the perspective of field, discourse semantics and lexicogrammar by way of establishing criteria for their classification.

Looking from above, I draw on the description of field presented in Martin (1992) as well as Doran and Martin (this volume). Field is re-articulated in Doran and Martin (this volume) as knowledge constructed from two complementary perspectives: a static perspective concerning classification and composition among items and a dynamic perspective considering the unfolding of activities as either *expectancy* or *implication* activities (i.e. as one activity 'expecting' what would follow or as one activity 'implicating' what must follow). This distinction resonates with previous descriptions of activity sequences and implication sequences (Martin 1992). Doran and Martin also consider qualitative and spatio-temporal properties potentially associated with activities and items.

Looking from around, I consider the interaction between entities and other discourse semantic systems, including in particular the interpersonal system APPRAISAL (Martin and White 2005) and the textual system IDENTIFICATION (Martin 1992). Due to space limitations, familiarity with these systems is assumed (see Martin 1992, Martin and White 2005 for details). Within discourse semantics, I consider meanings in terms of how entities are differentiated from and related to meanings construed by **figures** and **sequences** of figures (Hao 2015; cf. Halliday and Matthiessen 1999).

Looking from below, of particular concern are the grammatical realizations of entities at both clause and group ranks. For this analysis Halliday and Matthiessen's functional grammar of English (2014) is assumed.

It is important to note that in order to clearly distinguish meanings at different strata, in what follows, the terms 'taxonomy' and 'taxonomic relations' are used *only* to refer to meanings at the level of field; the discourse semantic resources construing taxonomies will be introduced, including **entities, co-elaborations** and **dimensions**. Figure 6.4 illustrates distinguishing terminologies across strata that we will attend to in the following discussion.

For the purpose of illustration, I present here in Table 6.4 the relevant entity types. The initial identification of entity types in Hao (2015, 2020b) includes broader choices – i.e. source, thing, activity, semiotic, place and time entities. This chapter focuses only on thing entities and activity entities (and their subtypes). It is important to note that it is not the naming of a category that is important, but rather the *criteria* the naming is based on. Each of the distinctions in Table 6.4 will be explored in the following sections.

Thing entity vs. activity entity

From the static perspective at the level of field, a scientific item (e.g. *a prokaryotic cell*) can be named through a discourse semantic entity, such as *prokaryotes* in (1).

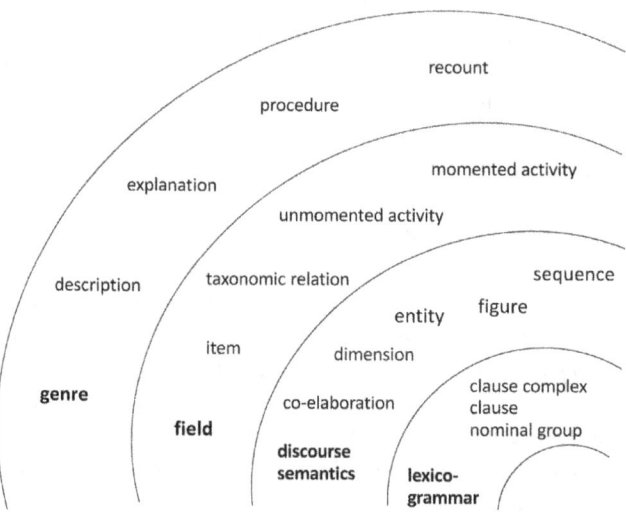

FIGURE 6.4 Distinctive terminologies across strata

TABLE 6.4 Entity types in undergraduate biology texts (c.f. Hao 2015, 2020b)

Types of entity			Examples
thing	instrumental	ostensively defined	*plate, container, microscope, pipette*
		linguistically defined: trained gaze	*glycerol medium, sodium carbonate*
	observational/ling. defined	trained gaze	*insects, gut, sea urchin, herbivore*
		tech enhanced gaze	*fungal spore, eukaryotic cells,*
		inferable	*pathogen, enzyme, cytoplasm*
activity	enacted/ostensively defined		*experiment, method, treatment*
	observed/ling. defined	trained gaze	*maceration, peristaltic movement*
		tech enhanced gaze	*germination, fungal spore dispersal*
		inferable	*enzymic digestion, gene expression*

The item can be potentially classified (e.g. *prokaryotes* vs. *eukaryotes*) or de/composed (e.g. *the plasma membrane* and *cytoplasms* are parts of *prokaryotes*).

(1) Prokaryotes [thing entity] were seen.

From a dynamic perspective, phenomena are considered in terms of activities. Activities can be either momented or unmomented. When an activity is unmomented, it can be realized through a figure such as in (2). Here a figure construing an activity is referred to as an **occurrence figure**, which is realized grammatically through a material process.

(2) A cell engulfs food.

When activities are momented into a series of activities, they can be realized through a sequence of figures in the discourse, as shown in (3).

(3) *After* a cell engulfs food by phagocytosis or pinocytosis,
     newly formed food vacuoles fuse with lysosomes.
     And then this mixes the food with the enzymes... (Reece *et al.* 2011: 880)

The figures in (3) are connected through temporal **connexions** (cf. Martin 1992) (*after, and then*).[3] The connexions are realized both within and across sentence boundaries. The connexion *after* is realized between the clauses in a clause complex, whereas the connexion *And then* is realized between two sentences. Connexions between figures can also be implicit. For example, the connexions in (3) can be realized as in (4), where the relation needs to be abduced from the surrounding discourse.

(4) A cell engulfs food by phagocytosis or pinocytosis.
     (and then) Newly formed food vacuoles fuse with lysosomes.
     (and then) This mixes the food with the enzymes.

In addition to figures and sequences, language allows activities to be construed in a static way. To illustrate this, we consider example (5), which is selected from the co-text of excerpt (3). Here a term *intracellular digestion* is used to name the momented activities realized in (3).

(5)   The hydrolysis of food inside vacuoles is called <u>intracellular digestions</u>. (Reece *et al.* 2011: 880)

By naming the momented activities through *intracellular digestion*, language enables the field activities to be encapsulated into one term. This language resource is coined as an **activity entity**, distinguished from the **thing entities** naming items such as *Prokaryotes*. Both thing entities and activity entities can enter into lexical cohesion in the discourse. Based on this discourse semantic understanding, Doran and Martin (this volume) suggest that activity entities allow field activities to be 're-construed' from a static perspective, becoming 'itemized activities'. Like items, an itemized activity can enter into classification and composition taxonomies. In example (6), the activity entities (underlined) construe part/whole composition taxonomic relations among itemized activities – i.e. *food processing* is the whole of four different parts.

(6)   The main stages of <u>food processing</u> are <u>ingestion</u>, <u>digestion</u>, <u>absorption</u>, and <u>elimination</u>. (Reece *et al.* 2011: 880)

Figure 6.5 consolidates the meaning choices across the strata and their typical interstratal relations.

At the lexicogrammatical level (looking from 'below'), the realization of activity entities can often take the form of nominalization. For example, *digestion* is a

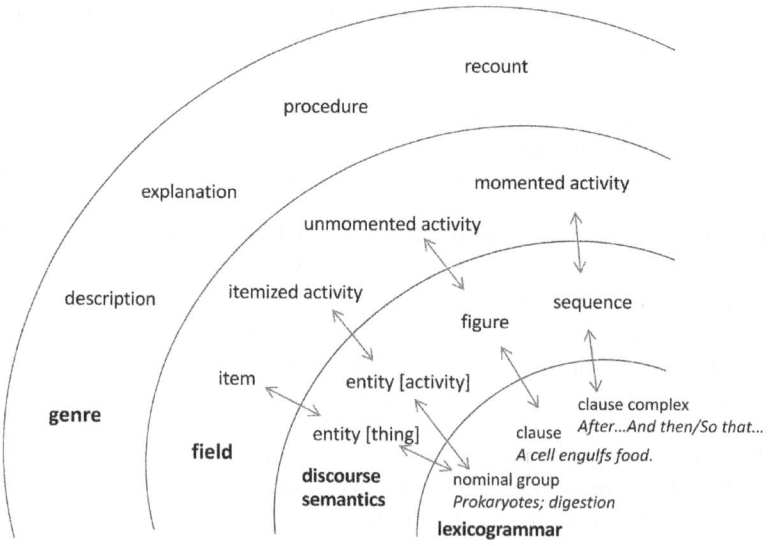

FIGURE 6.5   Choices across strata

nominalization derived from the verb *digest*. However, this is not always the case. Activity entities used in different fields can also draw on different lexical resources. This will be elaborated later when different types of activity entities are concerned.

Thing entities: instrumental vs. observational

Among thing entities, a further distinction can be made between *instrumental* thing entities, referring to items such as tools and apparatuses that are used in the laboratory during experiments (e.g. *pipette, glycerol medium*) and *observational* thing entities referring to biological items that are observed during laboratory experiments (e.g. *prokaryotic organism*). This distinction recognizes that the object of study and 'doing biology' (Hood 2010) are two different sets of disciplinary knowledge which implicate different sets of taxonomies.

To distinguish instrumental and observational thing entities, we can begin by looking from 'around' in the discourse semantics. Of particular relevance are the subtypes of appreciation in the ATTITUDE system (Martin and White 2005; Hao and Humphrey 2012), including valuation concerning the worthiness; composition concerning the organization; and reaction orienting to the significance. The distinction among these three types of attitude resonates with type of mental processes; that is to say, valuation is motivated by cognition ('thinking'), composition by perception ('seeing'), and reaction by affection ('feeling') (Martin and White 2005: 57). In the scientific discourse, perception of composition tends to be associated with measurable properties (cf. Hao and Humphrey 2012).

The two types of thing entities, instrumental and observational types, have distinctive ways of interacting with interpersonal meanings. Observational entities are typically evaluated through attitudinal choices of valuation, such as *play a crucial role* in (7); instrumental entities on the other hand are usually evaluated through composition such as *imprecise* in (8).

(7) Mandibles [observational entity] **play a crucial role** in the digestive process of the locust.
(8) The balance [instrumental entity] used (in the experiment) was **imprecise**.

In addition, from a textual perspective, observational and instrumental entities are identified differently in students' experimental reports. While both observational and instrumental entities are identified through specific reference, only instrumental entities are presumed exophorically, by referring to items 'outside' the text in the experimental setting – such as the instrumental entity *balance* in (8) (for identification resources, see Martin 1992, Martin and Rose 2007).

Grammatically, the different participant roles through which instrumental and observational entities are realized further confirm their opposition. In the students' experimental reports, instrumental entities are typically realized by a Goal in a material process, with people entities realized as Actors (typically implicitly in receptive clauses, as in (9)); observational things on the other hand tend to be realized through a Phenomena in mental processes, as in (10).

(9) In this experiment a <u>Finnpipette</u> [instrumental entity, Goal] was calibrated [material Process] (by <u>us</u> [people entity, Actor]).

(10) Both <u>prokaryotic and eukaryotic organisms</u> [observational entity, Phenomenon] were seen [mental Process] (by me/us [people entity, Senser]).

In addition to Participant roles, instrumental entities are also often realized in Circumstances, as either location (e.g. *container* in (11)) or manner (e.g. *sodium carbonate* in (12)).

(11) Set amount of water was pipetted (by us) into a <u>container</u> [instrumental entity, Location].

(12) The reaction was stopped (by us) with <u>sodium carbonate</u> (6.9mM) [instrumental entity, Manner].

The distinction between instrumental and observational entities is sensitive to the different activities they enter into, which leads us to distinguish between two types of activity entities.

Activity entities: enacted vs. observed activity entities

From the perspective of field, activity entities are further categorized through their relationship with types of activities – i.e. expectancy activity vs implication activity (Martin and Doran this volume). Expectancy activities in undergraduate biology are realized in the discourse through an *enacted* activity entity, including those referring to scientists' experimental methods, as those underlined in (13). Implication activities, typically those of the naturally occurring scientific phenomena, are realized through *observed* activity entities, such as *digestion* and *ingestion* in (14). Both activity entity types allow field activities to enter in the discourse without being momented.

(13) In this experiment a Finnpipette was calibrated, using three <u>methods</u>– <u>weight-of-water</u>, <u>spectrophotometry</u> and <u>radioactivity</u> [enacted activity entities].

(14) The viability of fungal spores may be determined by the effect of the physical and chemical processes involved in <u>ingestion</u> and <u>digestion</u> [observed activity entities].

In discourse analysis, distinguishing enacted and observed activity entities requires identifying their agnate construal as sequences of figures. Sequences agnate to enacted activity entities tend to be found in procedures such as laboratory manuals and recounts such as those named as 'Method' in student's experimental report (e.g. excerpt (15)).

(15) Set amounts of dye were pipetted into 1mL cuvettes,
　　　And water was added
　　　to give a total volume of 1mL.
　　　Each solution was mixed,
　　　and absorbances were read,
　　　using a spectrophotometer.

In contrast, sequences agnate to observed activity entities typically occur in explanations. For instance, the temporal sequencing of figures in excerpt (3), reproduced here in (16), agnates to the activity entity *intracellular digestion*, a type of *digestion*.

(16) After a cell engulfs food by phagocytosis or pinocytosis,
   newly formed food vacuoles fuse with lysosomes.
   And then this mixes the food with the enzymes…. (Reece *et al.* 2011: 880)

When recognizing sequences in the discourse, the difference between expectancy and implication activities is reflected in different logical connexions used in the sequence. Both temporal and causal connexions can be found; however, the connexions are used differently for construing different activity types. We consider first an example of expectancy activity, which can be itemized through an enacted activity entity, *spectrophotometry* in (13), and momented through a sequence in excerpt (17). This sequence involves explicit and implicit temporal [successive] connexions (italicized) between figures (i.e. *and then*), and manner (*by (means of)*) or purpose (*in order to*); they indicate a facilitating relationship between the figures.

(17) Set amounts of dye were pipetted into 1mL cuvettes,
   And water was added
   (in order) to give a total volume of 1mL.
   (And then) Each solution was mixed,
   and absorbances were read,
   (by) using a spectrophotometer.

In contrast, the fact that implication activities are both 'chronological and logical' (Barthes 1975) provides a distinctive criterion for identifying sequences agnate to observed activity entities. This means that causal and temporal connexions between figures are usually interchangeable. This can be exemplified by comparing the temporal connexions in (16) and the paraphrased causal ones in (18).

(18) *If* a cell engulfs food by phagocytosis or pinocytosis,
   then newly formed food vacuoles fuse with lysosomes.
   So then this mixes the food with the enzymes…

A second set of recognition criteria between enacted and observed activity entities arises in the discourse through their interaction with interpersonal and textual meanings (i.e. a perspective from 'around'). Interpersonally, as exemplified in (19) and (20), both enacted and observed activity entities can be coupled with appreciation [valuation] (in bold).

(19) Symbiotic relationships [observed activity entity] between termites and cellulose-producing gut fungi may be **beneficial** to both insects and fungi.
(20) …making such a study [enacted activity entity] **ecologically realistic and important**.

However, in addition to valuation, enacted activity entities can also be evaluated through appreciation [composition], such as in (21), which appears not to occur with observed activity entities.

(21) This method [enacted activity entity] was **time consuming**.

Textually, both enacted and observed activity entities can be presented as generic meanings. However, their generic presentation tends to draw on different lexicogrammatical resources. Enacted activity entities involve the non-specific determiners *a/an*, as in *a biocontrol* in (22). But observed activity entities are mostly (if not always) represented without a determiner, e.g. *symbiosis* in (23).

(22) If entomopathogens are to be developed towards a biocontrol [enacted activity entity] …
(23) Members of the Chytridiomycota may be involved in symbiosis [enacted activity entity] with the Echinoidea.

In addition, enacted activity entities differ from the observed ones in that they can also be presumed by tracking a specific meaning in the preceding text – such as *this method* in (21), pointing back in the text where the method was recounted in Method stage. However, observed activity entities are not normally presumed.

Grammatically, while nominalization is a common lexicogrammatical realization of both enacted and observed activity entities, enacted activity entities do not always involve nominalizations – especially in more common sense (i.e. *party* and *game*) and specialized fields (e.g. *volley* and *overhead* in tennis (Martin 2017)). By contrast, observed activity entities appear regularly as nominalizations, often drawing on Latin and Greek etymology (White 1998) (e.g. *vaccination* from the Latin *vacca*, cow; *phagocytosis* from the Greek *phagein*, to eat). When activity entities are realized nominally, they can be confused with experiential metaphors, particularly realizing figures metaphorically through nominal groups (e.g. *ingestion* [activity entity] *is a stage of food processing* vs. *the ingestion* [experiential metaphor] *of algae was successful*). A key distinction between the two is that activity entities are generalizable across texts, whereas the use of experiential metaphor is specific to the unfolding of a particular text. For an in-depth discussion on the distinction between activity entities and experiential metaphors, see Hao (2020a).

Identifying enacted and observed activity entities has enabled us to reveal the discourse semantic resources construing field activities from a static perspective.

Before taking a further step, we take stock in Figure 6.6 of the oppositional choices of entities that have been so far identified.

Linguistically defined vs. ostensively defined entities

A further distinction simultaneous with the thing and activity entities distinction arises in terms of how these entities interact with textual meanings composing

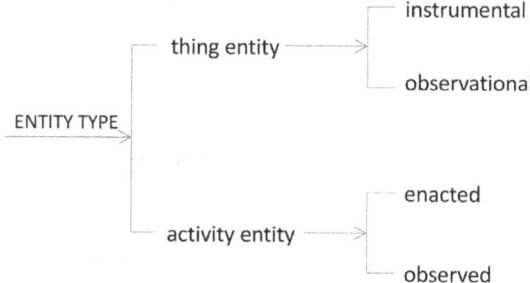

FIGURE 6.6 System of ENTITY TYPE (1)

mode. Based on studies in language development (Painter 1999), Martin suggests that meanings in more common sense fields (including everyday and specialized fields) are learned ostensively in spoken language as they are pointed out based on one or another dimension of sensory experience (to see, touch, taste, hear and/or smell) (2017: 117). By contrast, educational knowledge such as science relies heavily on learning through language in written modes, in institutional settings (Martin 2007: 40–41). This interaction of field with mode motivates a further entity distinction between those which are ostensively defined in relation to sensuous experience, and those which are linguistically defined through definitions in written language.

This opposition is reflected in the way entities are textually identified through choices in IDENTIFICATION. An ostensively defined entity used in an everyday discourse typically refers exophorically to a meaning 'outside' the text, as in example (24) (selected from Painter (1999: 84)).

(24) Child: What's **that**?
  Mother: **That** [exophoric pointing; Token] 's <u>my belly button</u> [specific; Value].

Through identification of this kind, an item enters into the discourse as an entity. Building on the naming of such instances, young children can subsequently develop an orientation to classes – e.g. *dolphins are mammals*, with both entities being 'named'.

In the written texts of undergraduate biology, identification is more complex. While some instances of instrumental things tend to combine with exophoric reference and implicate ostensively defined meanings (i.e. *the balance* in (7) above), the majority of entities are introduced and tracked within the text and they are typically provided with linguistic definitions in pedagogic materials.

Linguistic definitions explicitly establish relationships between entities. In the definition of *lysosome* below, *lysosome* is related to other entities including *membranous sac, hydrolytic enzymes* and *animal cell*.

(25) <u>A lysosome</u> [Token] is *a membranous sac of hydrolytic enzymes that an animal cell uses to digest all kinds of macromolecules* [Value] (Reece *et al.* 2011: 106)

In addition to linguistic definitions, presentations of 'how things look' in pedagogic texts rely heavily on multimodal resources, including photographs, illustrations and

▸ Kingdom Fungi
is defined in part
by the nutritional
mode of its members (such as this mushroom), which
absorb nutrients from outside their bodies.

FIGURE 6.7　Visual representation of an item in undergraduate textbook (Reece *et al.* 2014: 11)

diagrams. These images present instances of entities. For example, in Figure 6.7, the caption describes one of the distinctive features of an item introduced as a generic entity (*Kingdom Fungi*). The specific reference to the image – *this mushroom* – exemplifies the generic category.

Imagic exemplification of linguistically defined entities suggests a difference between characterizing phenomena through ostensive definition and doing so in terms of their tangibility. This will be the focus of the next section.

Tangibility of linguistically defined entities

The tangibility of linguistically defined entities allows us to identify three further types: trained gaze entities referring to items and activities can be observed with human senses, tech-enhanced gaze entities referring to items and activities whose observation requires technology, and inferable entities referring to items and activities whose physical existence is inferred rather than directly seen and observed.

The tangibility of linguistically entities can be revealed in terms of how they interact with images. A trained gaze entity, such as a *sea urchin*, is often represented in pedagogic texts through realistic photographs such as the image in Figure 6.8. This realistic image illustrates the tangible observable nature of trained gaze entities. The caption here maintains the description generically (e.g., *sand dollars, sea stars, marin animals, a sea urchin*).

Like trained gaze entities, tech-enhanced gaze entities also refer to relatively tangible items; but they need to be perceived with technology. For instance, we can observe the presence of *Chytrids,* a kind of *fungi*, using a microscope. Its presence can be realistically captured and visually represented, such as in Figure 6.9. This observation would have been impossible with our naked eyes. The image provides an instance of the entities which are presented generically in the caption (e.g., *Chytrids, multicellular, branched hyphae*).

Echinodermata (7,000 species)

Echinoderms, such as sand dollars, sea stars, and sea urchins, are marine animals in the deuterostome clade that are bilaterally symmetrical as larvae but not as adults. They move and feed by using a network of internal canals to pump water to different parts of their body (see Concept 33.5).

A sea urchin

FIGURE 6.8 A visual representation of a trained gaze entity – *sea urchin* (Reece *et al.* 2014: 683)

Chytrids (1,000 species)

In chytrids such as *Chytridium*, the globular fruiting body forms multicellular, branched hyphae (LM); other species are single-celled. Ubiquitous in lakes and soil, chytrids have flagellated spores and are thought to include some of the earliest fungal groups to diverge from other fungi.

Hyphae   25 μm

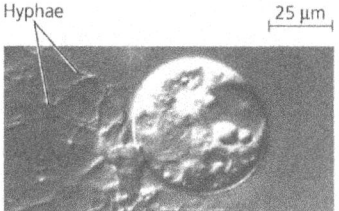

FIGURE 6.9 Visual representation of *Chytrids* (Reece *et al.* 2014: 655)

In contrast to the interaction with realistic imagic representation, inferable entities tend to be made 'visible' through illustrations and diagrams, which Kress and van Leeuwen (2006) refer to as analytical diagrams. In the visual representation in Figure 6.10, both the inferable activity entity *chemical reaction* and the inferable things involved (i.e. *glucose molecules, simpler molecules*) are represented through illustrations.

The interaction between entities and images suggests that different types of linguistically defined entities have different choices of visual representation.[4]

To summarize, we represent the entity types for scientific taxonomy building in Figure 6.11.

These entity types are helpful for thinking about the diversity of taxonomies in biological sciences. However, in order to fully understand taxonomy building including diversity as well as its depth, we need to further explore the relationships between entities of the same type.

Dimensionality

In this section we explore the depth of taxonomies by describing the relationships between entities of the same type. The resources for differentiating one entity from

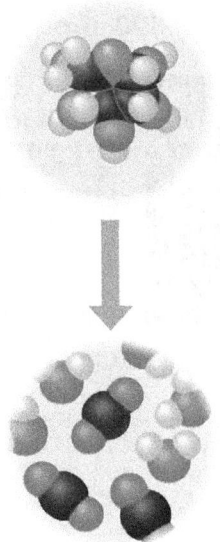

(c) Chemical reaction. In a cell, a glucose molecule is broken down into simpler molecules.

FIGURE 6.10 Visual representation of *chemical reaction* (Reece *et al.* 2014: 146)

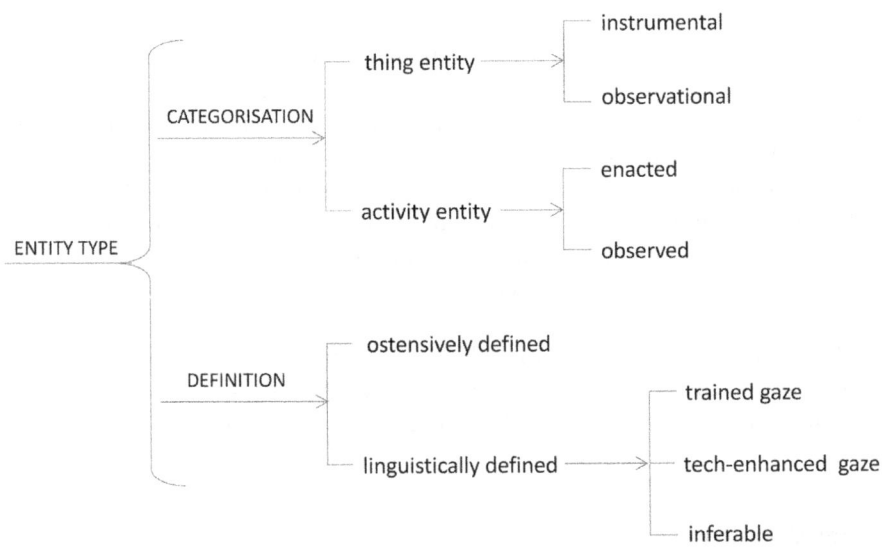

FIGURE 6.11 System of ENTITY TYPE (2)

another (of the same type) are conceptualized here as discourse semantic augmentation of entities, referred to as dimensions. Below are some examples of dimensions (in italics) augmenting entities (underlined).

> the *size* of the <u>cell</u>
> the *shape* of <u>sea urchin</u>
> the *kind* of <u>birds</u>
> *parts* of a <u>cell</u>

The system of dimensions, DIMENSIONALITY, is simultaneous with the system of ENTITY TYPES in the ENTITY system. In other words, all entities can be dimensioned. The more an entity is dimensioned, the more in-depth knowledge of the item is developed. The symbol '<' and '>' are used to annotate the augmenting relationship between entity and dimension, depending on their order in the text. The arrow points always towards the entity (e.g. *the size of > the <u>cell</u>; the <u>cell's</u> < size*).

Looking from above, dimensions allow properties and relationships among items and activities to be catalogued and named through language. Properties, which are typically realized through qualities in the discourse, such as *big* in (26), are 'itemized' (Martin and Doran this volume) through dimensions such as *size* in (27). Notice that Focus^Thing structure of a nominal group (e.g., *the size of the cell*; for Focus group analysis see Martin *et al.* 2010) provides a canonical grammatical realization of dimension>entity.

(26) The cells [entity] are **very small** [quality].
(27) The *size* [dimension] of > the <u>cell</u> [entity] was measured.

In building taxonomic relations, entities in the discourse are co-related with one another. As the text unfolds, each instance in the discourse is understood not in isolation but in relation to its superordinate, co-classes and subclasses, and to its whole, co-parts and parts. The relationship between entities is 'co-elaborative' (annotated as 'entity=entity') (e.g. *dolphins are mammals [dolphins = mammals]*), in the sense that two related entities both indicate the classification and composition with respect to each other. Co-elaboration can be either realized lexicogrammatically, through a range of grammatical resources, or abduced from the co-text. For example in (28), the co-elaboration between *B-galactosidase* and *enzyme* is realized through an elaborating nominal group, their further co-elaboration to *protein* is subsumed in the text.

(28) The activity of <u>proteins</u> can be controlled… In this experiment the activity of <u>B-galactosidase</u>, an <u>enzyme</u> which breaks down lactose, was studied.

While co-elaborations construe taxonomic relations as the text unfolds (see Hood and Hao this volume), dimensions can provide explicit naming of taxonomic relations established, such as *kind* in (29).

(29) enzyme [entity] is a *kind* [dimension] of > protein [entity].

TABLE 6.5 Realizing taxonomic relations through co-elaborations and dimensions

Co-elaboration without dimension	Co-elaboration with dimension
Thing^Qualifier	parts of > the sea urchins;
the <u>coelomic cavity</u> and <u>digestive tract</u> of = the <u>sea urchins</u>	
possessive Deictic^Thing	part of > rainforest;
<u>rainforest's</u> = <u>canopy</u>	
elaborating nominal group complex	a kind of >irregular urchin;
Chytridiomycota were present in the digestive system of	the superordinate of > Echinocardium
1 the irregular urchin, = _2 Echinocardium cordatum_	cordatum
relational identifying process	a kind of > enzyme;
<u>B-galactosidase</u> is = an <u>enzyme</u>	the superordinate of > B-galactosidase
which breaks down lactose.	
relational attributive process	a kind of > food source
<u>Glucose</u> is = a preferred simpler <u>food source.</u>	

Table 6.5 illustrates a range of co-elaborations construing taxonomic relations which can be potentially named with dimensions.

Before drawing perspectives from around and below, we present an overview of the types of dimension identified in undergraduate biology texts (Table 6.6) – categorized, structured, measured and perceived dimensions (cf. Hao 2015). Categorized dimensions offer a way of talking about how items are interrelated as superordinate to subtype, or class to instance; structured dimensions include resources for relating items as parts to wholes; measured dimensions provide ways of talking about properties which are otherwise realized in the discourse through qualities (e.g., _high, small, long_) and quantities (e.g., 5 meters); and perceived dimensions name properties based on our sensory experience (to see, smell, touch, hear, etc.).

When looking 'from around' at the level of discourse semantics, entity and dimension as a whole interact with other discourse semantic meanings. For example in (30), _Members of Chytridiomycota_ as a whole is an evaluated target (_ecologically important_). Similarly in (31) what is being _measured and recorded_ is 'measured dimension>entity' as a whole.

(30) _Members_ of > <u>Chytridiomycota</u> [categorized dimension>observational entity] are **ecologically important** [valuation].

TABLE 6.6 Dimensionality of entities (c.f. Hao 2015)

Types of dimension	Examples
categorized	type, kind, category, species…
structured	part, component, layer, level, stage…
measured	size, weight, height, length…
perceived	color, shape, texture, smell….

(31) The *weight* of > <u>the water</u> [measured dimension>instrumental entity] was measured and recorded.

Instances of dimensions can sometimes appear with an elided entity, such as *the weight was measured*. Such instances occur only after the entity has been introduced in the preceding text. As exemplified in (32), the omitted entity *DNA*, which is elaborated with a measured dimension *quantity*, can be recovered based on its presentation in the previous clause. The recovered choices are annotated with superscript.

(32) Isolating <u>DNA</u> and determining *the quantity* > ^{of <u>DNA</u>} extracted...

It is also possible to elide a dimension. As exemplified in (33) and (34), the elided perceived dimension *shape* and categorized dimension *kind/instance* can be readily inferred, shown in the superscript.

(33) ^{The *shape* of >} the <u>sea urchin</u> is irregular.
(34) This ^{*kind/instance* of >} <u>irregular urchin</u> is called <u>Echinocardium cordatum.</u>

Given the diverse range of linguistically defined entities in biology texts, dimensions of linguistically defined entities are also found to be 'field-specific' in biology. Scientists have developed 'field-specific' ways of naming the relationships among items of organisms, such as *domain, kingdom, phylum, class, order, family, genus* and *species*. Field-specific dimensions typically also require linguistic definition. For example, *species* is defined in the textbook as following: *each of these forms of life* (i.e. *plants, animals, fungi* and *bacteria* in the proceeding text) *is called a species* (Reece *et al.* 2011: 4).

A further characteristic of field-specific dimensions is that they may originate from grammatical metaphors. This is particularly the case in biology for measured dimensions, as part of the distillation of technicality. For example, a number of field-specific measured dimensions in (35) such as *activity* are associated with inferable entities (*proteins, B-galactosidase*):

(35) The *activity* of > <u>proteins</u> can be controlled... In this experiment the *activity* of > <u>B-galactosidase,</u> an enzyme which breaks down lactose, was studied.

The nominalization *activity* here is derived from the adjective *active*. However, instead of representing an experiential metaphor, the term *activity* used here is defined in the textbook as *how efficiently the enzyme functions* and the measurement of *activity (of protein/enzyme)* is based on a mathematical equation of quantification: *Activity of enzyme = moles of substrate concerted per unit time = rate × reaction volume.*

Grammatically, dimensions are realized either canonically through a Focus group in Focus^Thing structure, or through a Thing when the entity is elided, or being rendered as other nominal group structures as outlined in Table 6.7.

TABLE 6.7 Grammatical realizations of dimensions in biology texts

Lexicogrammatical realizations of dimensions	Examples
Thing	*type; parts; shape; weight*
Focus^Thing	*a kind of >sea urchin;*
	members of > the geofungi
Classifier^Thing	*regular sea urchin < species*
Deictic^Thing	*sea urchin's < shape*
Thing^Qualifier	*mandibles < of different sizes*

I have now illustrated the choices for discourse semantic entities and their associated dimensions in undergraduate biology texts. The description identifies resources that connect taxonomies of items and activities at the level of field with a range of realizations at the level of lexicogrammar. In the next section, I demonstrate how this description enables a thorough exploration of taxonomies demonstrated in an excerpt of research report produced at the final undergraduate year.

Analyzing taxonomy in a research report

The selected text for the illustrative analysis is a student research report produced at the final year of undergraduate coursework. The text reports on an investigation of insects interacting with fungal spores. As an illustration, I will focus on the Abstract previewing the report. The excerpt (36) presents this annotated text. The entities are underlined; the dimensions are in italics; and the elided entities, dimensions and their substitutions are made explicit and annotated as superscripts.

(36) The *viability* of > fungal spores after ingestion and passage through the gastrointestinal tract of an insect may be determined by the effect of physical and chemical processes involved in the ingestion and digestion of food. In particular, mandibular maceration could damage fungal spore *< integrity* and result in spores losing their [(fungal spores')] *< viability*. The purpose of this study was to determine whether the *size* of > fungal spores was a factor in affecting spore *< viability* after passage through the gastrointestinal tract of the Australian plague locust, Chortichocetes terminifera. The effect of spore *< size* on *viability* > [of spore] was tested using five *genera* > of dung fungi – Absidia, Isaria, Penicillin, Phycomycetes and Podospora – whose spores were fed to either second or fifth instar C. terminifera. Absidia[spores], Isaria[spores] and Penicillin spores were recovered from both the faecal and gut *< samples* from second[instars] and fifth instars, following feeding by C. terminifera on wheat inoculated with fungal spores. Phycomyces was not recovered from faecal material obtained from second instars but was present in all other *samples* > [of gut]. Podospora spores, 20um in diameter, were not recovered from any of the *samples* > [of gut].

The analysis outlines the complex taxonomies construed in this text both in terms of their diversity and depth. The discussion breaks down the analysis based on different entity types.

To begin, the text demonstrates significant use of thing entities, including both tech-enhanced things and trained gaze things. Tech-enhanced realizations refer mostly to the item *dung fungus* and its part *fungal spore*. The entities are co-elaborated in relation to one another. For example, *dung fungi* as a superordinate is co-elaborated by the subtypes *Absidia, Isaria, Penicillin* and *Phycomycetes*. The classification is named through the categorized dimension *genera*. The co-elaborations associated with *dung fungus* construe the taxonomy displayed in Figure 6.12. The diagram also includes the item *Zygomyce* appearing in the subsequent text (in Discussion section) of the research report. The items unmentioned in the report are left implicit in the diagram.

In addition to the categorized dimension, measured dimensions augmenting the entity *fungal spore* are also salient in this text, including both field-neutral measurements (e.g., *size*) and field-specific ones (e.g., *integrity, viability*). These measured dimensions construe properties of the item from a static perspective. The itemized properties allow for the re-construal of the properties that are typically realized through qualities and quantities in the discourse. In example (37), selected from the Discussion of the report, a property the item *spores* is realized through a quality *larger* and a quantity *14-20um*. In the Abstract section the property is itemized through the measured dimension *size,* as shown in (38).

(37) The larger spores of Podospora (14–20um) did not retain their viability.
(38) The purpose of this study was to determine whether the *size* of > fungal spores
     was a factor in affecting spore viability

The multiple dimensions augmenting *fungal spore* indicate that establishing properties of the item is a prominent feature of this text, reflecting an in-depth knowledge of the item.

The other salient thing type in this excerpt is trained gaze things (e.g., *insects, gut* and *Australian plague locust*). These entities are co-elaborated around the entity *locust*, construing both its classification and composition. For example, *Chortichocetes terminifera* as a kind of *locust* is indicated by an elaborating nominal group complex

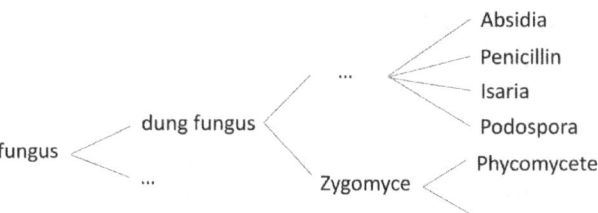

FIGURE 6.12   Categorization of *dung fungus*

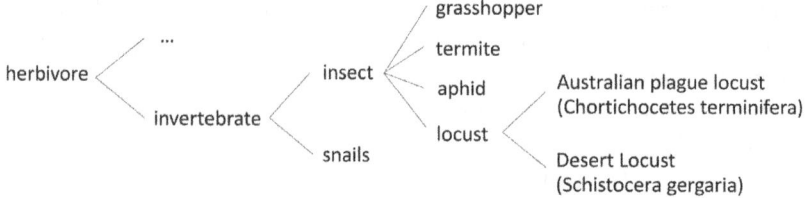

FIGURE 6.13 Categorization in relation to *locust*

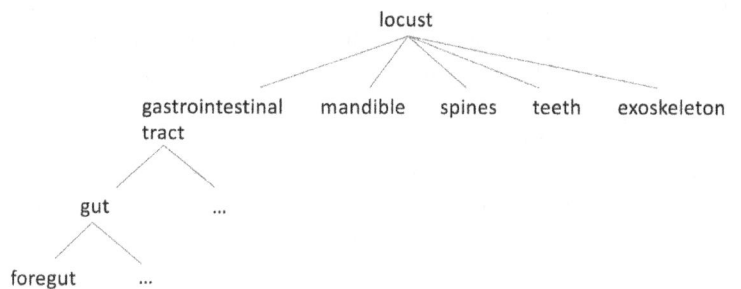

FIGURE 6.14 Structure of *locust*

(*Australian plague locust*, = *Chortichocetes terminifera*). Its further elaboration to higher level category (i.e. *locust* = *insect*) is subsumed in the text. Similarly, the part/whole co-elaboration between *the gastrointestinal tract* and *locust* is realized through a Thing^Qualifier structure (e.g., *the gastrointestinal tract* = *of an insect*). While an instance of categorized dimension naming classification was found (e.g., *sample* relating to class to instance), no structured dimensions naming composition were found.

Figure 6.13 and Figure 6.14 provide an overview of the classification and composition in relation to *locust* established throughout the whole text of the research report.

Along with the tech-enhanced things, corresponding activity entities (e.g., *mandibular maceration, ingestion*) were also found. The activity entities are co-elaborated by connecting activities both at different tiers (e.g., *physical and chemical processes* = *ingestion/digestion*), and at the same tier (e.g., *digestion* = *ingestion*). However, the co-elaborations are not named as dimensions (e.g., *a stage of digestion*; *stages of food processing*), indicating that classification and composition of activities are not the focus of the study.

Few instances of enacted entities were identified in the excerpt. This suggests that its object of study focuses on the biological knowledge (i.e. 'what is observed'), rather than research methodology (i.e. 'how to observe').

In summary, the preview of research report reveals relative broad and deep taxonomies, referring to tangible and observable items and activities. The most elaborated and augmented entities are those of trained gaze and tech-enhanced ones. This suggests that demonstrating static knowledge of items and activities at different levels of tangibility is an important task in this text.

Onward and outward

In this chapter I have illustrated a discourse semantic framework for analyzing taxonomies in undergraduate biology. Importantly, the chapter illustrates a 'tri-stratal' method for describing discourse semantic meanings, particularly entity types and their dimensions. With this method, highly technical meanings of entities and dimensions can be made transparent. The approach can be used to explore how language construes taxonomies across disciplinary fields.

As far as undergraduate biology is concerned, the analysis of taxonomies in the excerpt of a student's research report reveals a diverse range of entities referring to relative tangible items (observed either through naked eyes or through microscope). This suggests that being able to establish taxonomies of items and activities at different levels of tangibility is significant for the knowledge development of biological sciences.

Dialogue with LCT, in particular with respect to knowledge building, has encouraged the development of further work on the 'language' of disciplinary knowledge. SFL's perspective on entities and dimensions can be related to many aspects of knowledge building in LCT, including the 'epistemic-semantic density' concepts presented in Maton and Doran (2017a, 2017b). This paper is conceived as a contribution to further productive interdisciplinary conversation.

Notes

1 Grammatical metaphors have been distinguished as experiential, interpersonal and logical metaphors (Halliday 1994, Martin 2008). This chapter will only be concerned with experiential metaphors.
2 Note that the use of Focus group here follows the terminology in Martin, Matthiessen and Painter (2010). Focus group is referred to as Pre-Deictic in Martin (1992).
3 Connexion is used to substitute Martin's (1992) discourse semantic 'conjunction'. This reserves 'conjunction' as a grammatical term as used in Halliday and Matthiessen (2014).
4 It is important to note that while some items are usually inferred in biological experiments, in chemistry they can be shown through technology such as the sophisticated imaging instrument PET scanner, which can monitor chemical processes. In other words, the distinction between a tech-enhanced gaze entity and an inferable entity in a field depends on the different technologies deployed.

References

Barthes, R. (1975) 'Introduction to the structural analysis of narrative', *New Literary History*, 6(2): 237–72.
Bernstein, B. (2000) *Pedagogy, Symbolic Control and Identity,* London: Rowman and Littlefield.
Halliday, M. A. K. (1994) *An Introduction to Functional Grammar*, London: Edward Arnold.
Halliday, M. A. K. (1998a) 'Things and relations: Regrammaticizing experience as technical knowledge', in J. R. Martin and R. Veel (eds) *Reading Science,* London: Routledge, 185–236.
Halliday, M. A. K. (1998b) 'Language and knowledge: The "unpacking" of text', in J. Webster (Ed.) *The Language of Science*, London: Continuum, 24–48.

Halliday, M. A. K. and Matthiessen, C. M. I. M. (1999) *Construing Experience Through Meaning: A language-based approach to cognition*, London: Cassell.

Halliday, M. A. K. and Martin, J. R. (1993) *Writing Science: Literacy and discursive power*, London: Falmer Press.

Hao, J. (2015) '*Construing Biology: An Ideational Perspective*', unpublished PhD thesis, University of Sydney.

Hao, J. (2018) 'Reconsidering "cause inside the clause" in scientific discourse – from a discourse semantic perspective in systemic functional linguistics', *Text and Talk*, 38(5): 520–50.

Hao, J. (2020a) 'Nominalisation in scientific English: A tristratal perspective', *Functions of Language*, 27(2): 143–73.

Hao, J. (2020b) *Analysing Scientific Discourse from a Systemic Functional Linguistic Perspective: A Framework for Exploring Knowledge Building in Biology*, London: Routledge.

Hao, J., and Humphrey, S. (2012) 'The role of 'coupling' in biological experimental reports', *Linguistics and the Human Sciences*, 5(2): 169–94.

Hood, S. (2010) *Appraising Research: Evaluation in Academic Writing*, New York: Palgrave Macmillan.

Kress, G. R. and van Leeuwen, T. (2006) *Reading Images: The grammar of visual design*, New York and London: Routledge.

Martin, J. R. (1992) *English Text: System and Structure*, Amsterdam: Benjamins

Martin, J. R. (1993) Literacy in science: learning to handle text as technology, in M. A. K. Halliday and J. R. Martin (eds) *Writing Science: Literacy and discursive power*, London: The Falmer Press, 166–202.

Martin, J. R. (1997) Analysing genre: functional parameters, in F. Christie and J. R. Martin (eds) *Genre and institutions: Social processes in the workplace and school*, London and New York: Continuum, 3–39.

Martin, J. R. (2007) Construing knowledge: a functional linguistic perspective, in F. Christie and J. R. Martin (eds) *Language, Knowledge and Pedagogy: Functional linguistic and sociological perspectives*, London and New York: Continuum, 34–64.

Martin, J. R. (2008) 'Incongruent and proud: De-vilifying 'nominalization''', *Discourse and Society*, 19(6): 801–10.

Martin, J. R. (2013) *Systemic Functional Grammar: A Next Step into the Theory – Axial Relations*, Beijing: Higher Education Press.

Martin, J. R. (2017) 'Revisiting field: Specialized knowledge in Ancient History and Biology secondary school discourse', *Onomázein Special Issue*, March: 111–48.

Martin, J. R. and Maton, K. (2013) (eds) 'Cumulative knowledge-building in secondary schooling', *Linguistics and Education* 24(1), 1–74.

Martin, J.R. and Maton, K. (2017) 'Systemic functional linguistics and Legitimation Code Theory on education: Rethinking field and knowledge structure', *Onomázein Special Issue*, March: 12–45.

Martin, J. R. and Rose, D. (2007) *Working with Discourse: Meaning beyond the clause*, New York: Continuum.

Martin, J. R. and Rose, D. (2008) *Genre Relations: Mapping culture*, London: Equinox.

Martin, J. R. and White, P. R. R. (2005) *The Language of Evaluation: Appraisal in English*, London: Palgrave.

Martin, J. R., Matthiessen, C. M. I. M. and Painter, C. (2010) *Deploying Functional Grammar*, Beijing: The Commercial Press.

Maton, K. (2014) *Knowledge and Knowers: Towards a Realist Sociology of Education*, London: Routledge.

Maton, K. and Chen, R. T.-H. (2020) Specialization codes: Knowledge, knowers and student success, in J. R. Martin, K. Maton and Y. J. Doran (eds) *Accessing Academic Discourse,* London: Routledge, 35–58.

Maton, K. (2020) 'Semantic waves: Context, complexity and academic discourse', in J. R. Martin, K. Maton and Y. J. Doran (eds) *Accessing Academic Discourse,* London: Routledge, 59–85.

Maton, K. and Doran, Y. J. (2017a) 'Semantic density: A translation device for revealing complexity of knowledge practices in discourse, part 1 – wording', *Onomázein Special Issue,* March: 46–76.

Maton, K. and Doran, Y. J. (2017b) 'Condensation: A translation device for revealing complexity of knowledge practices in discourse, part 2 – Clausing and sequencing', *Onomázein Special Issue,* March: 77–110.

Maton, K., Hood, S. and Shay, S. (eds) (2016) *Knowledge-building: Educational studies in Legitimation Code Theory,* London: Routledge.

Matthiessen, C. M. I. M. and Halliday, M. A. K. (1997/2009) *Systemic Functional Grammar: A first step into the theory,* Beijing: Higher Education Press.

Painter, C. (1999) *Learning through Language in Early Childhood,* London: Cassell

Reece, J. B., Urry, L. A., Cain, M. L., Wasserman, S. A., Minorsky, P. V., and Jackon, R. B. (2011) *Campbell Biology* (9th Edition), San Francisco: Pearson.

Reece, J. B., Urry, L. A., Cain, M. L., Wasserman, S. A., Minorsky, P. V. and Jackon, R. B. (2014) *Campbell Biology* (10th Edition), San Francisco: Pearson.

White, P. (1998) 'Extended reality, proto-nouns and the vernacular: Distinguishing the technological from the scientific', in J. R. Martin and R. Veel (eds) *Reading Science,* London: Routledge, 167–97.

Wignell, P., Martin, J. R. and Eggins, S. (1993) 'The discourse of geography: Ordering and explaining the experiential world', in M. A. K. Halliday & J. R. Martin (eds) *Writing Science,* London: Falmer, 151–83.

7

MULTIMODAL KNOWLEDGE

Using language, mathematics and images in physics

Y. J. Doran

Introduction

What knowledge do students need in order to be successful in science? Such a question pushes at the heart of educational programmes that aim to develop a comprehensive pedagogy that reaches across disciplines. In one sense, the answer is simple. The knowledge students need is specified in the syllabus, detailed in the textbook and explained in the classroom. From this perspective it is simply a matter of reading the textbook, listening to the teacher and writing down the knowledge in the exam or in an assignment. However, few in educational research would argue for such a simplistic model. For one, it is clear from decades of research into literacy across the curriculum that the processes of reading, listening and writing are far from unproblematic. From an educational linguistic perspective, there is a range of highly specific literacies at stake; students must be able to read and write a wide range of scientific genres, and interpret and make use of highly intricate scientific language (Martin 1985, Lemke 1990, Rose et al. 1992, Halliday and Martin 1993, Christie and Martin 1997, Martin and Veel 1998, Unsworth 1997, 2001a, Halliday 2004, Martin and Rose 2008, Martin and Doran 2015, Hao 2020). In the classroom they need to be able to listen and interpret what the teacher is saying, engage with them in dialogue and successfully reconstrue scientific meanings at the teacher's bidding (Rose 2004, 2014, 2020, chapter 11 of this volume, Christie 2002, Rose and Martin 2012). These literacy demands involve not just language but also extend to the multiliteracies inherent in science schooling – where language, mathematics, images, specialized symbolic formulae, animations and demonstration apparatus all need to be 'read' as one and reorganized where necessary in assignments and exams (Lemke 1998, 2003, Kress et al. 2001, Unsworth 2001b, O'Halloran 2005, Parodi 2012, Doran 2017, 2018, 2019, Doran and Martin, Chapter 5 of this volume). To learn the knowledge of a discipline, one must be able to grasp the way it is construed through language and other semiosis; and to show you have the knowledge, you must be able to marshal these specialized linguistic and semiotic resources to reconstrue these meanings.

Such mastery of a wide range of literacy demands is in many ways the crux of many issues in learning how to do science. But simply being able to read and write scientific language and the particular text types needed across the curriculum is not enough. Students need to understand where and when each particular text type, semiotic resource or particular linguistic resource is appropriate. They need to be able to interpret new situations and new demands and organize the meanings they have learnt in an appropriate manner. That is, they need to understand the principles underpinning the selection and application of particular meanings and why they are used. At stake here is a way of seeing the world. Students must develop a scientific gaze that allows them, among other things, to shift between the knowledge of the empirical world and that of abstract theory, and between everyday understandings and technical conceptions. They need to be able to make connections between phenomena that at first glance may seem disparate and carry out specialized procedures and protocols to investigate prescribed phenomena. And they must be able to marshal particular literacies to do this at the appropriate time.

This chapter explores how knowledge is organized multimodally in science, focusing in particular on physics as it is taught in schooling. It will explore two main concerns: the technical meanings construed through the key resources of language, image and mathematics, and the underlying principles that organize a scientific gaze and enable students to understand when to use particular technical meanings and semiotic resources. To do this, it will view the meanings made from two perspectives. First it will consider scientific meanings through the register variable *field* from Systemic Functional Linguistics (SFL) (Doran and Martin, chapter 5 of this volume). Second it will explore these meanings through the sociological framework of Legitimation Code Theory (LCT), which is being widely used alongside SFL (Maton and Doran 2017a, Martin et al., 2020), specifically its dimension of Semantics (Maton 2014, 2020).

The perspective from field in SFL will explore how particular technical meanings are organized by different resources – language, mathematics and image. Here we will be concerned with whether technical meanings are presented *statically* as set of items positioned in taxonomies (either of classification – type/subtype, or of composition – part/whole), or whether they are presented *dynamically* as series of events (known as activities) given in more or less detail. In addition, we will be concerned with whether these items or activities involve particular properties that can be measured and to what degree all of these meanings are placed in large, interdependent networks of meaning. Each dimension of field will be introduced in more detail as they become relevant, but this analysis will show that particular semiotic resources tend to specialize in particular ways of organizing meaning, while backgrounding others. In this sense, this chapter will show that the organization of content in many disciplines is inherently multimodal, as particular components of its technical array tend to be given full expression through particular semiotic resources.

Complementing this view from SFL we will also consider knowledge in physics from the perspective of Semantics in LCT (Maton 2014, 2020). This will give an insight into two organizing principles that characterize knowledge practices. The first is known as *semantic gravity* (SG), which conceptualizes the degree of context-dependence of meaning. Stronger semantic gravity (SG+) indicates meanings are more dependent on their context; weaker semantic gravity (SG−) indicates meanings are less dependent on their context. For example, in a primary school physics text that we return to below, the concepts of pushing and pulling are introduced by listing a series of examples, such as *Michael pushes the keys to a tune*, which refers to an image of a child playing a keyboard (Riley 2001). In this case, we would say that the *pushes* here involves relatively strong semantic gravity (SG+) as it is dependent on a particular instance of pushing − it has relatively high context-dependence. In contrast, later on in the book, the text generalizes in order to introduce the concept of *force*, using the sentence *A push is a force*. Here *push* does not make reference to any particular instance of pushing, rather can be applied to any number of instances. In this case, we can describe *push* as indicating weaker semantic gravity (SG−) than the first instance as it has less context-dependence. Shifts in semantic gravity are crucial to the organization of knowledge in physics (and indeed all sciences) and are organized through language, mathematics and image.

The second organizing principle at the centre of Semantics in LCT is *semantic density* (SD). Semantic density refers to the complexity of meaning. Stronger semantic density (SD+) indicates more complex meanings; weaker semantic density (SD−) indicates less complexity. In the sentence *a push is a force*, the term *force* is a technical term in physics. As students move through the years, this term becomes more centrally integrated into the vast network of meaning organizing physics knowledge. For example, it underpins Newtown's laws of motion, which organize the field of mechanics; it underpins different types of field in both nuclear physics and in relativity; and it forms a variable within innumerable mathematical equations at all levels. In this sense, *force* exhibits relatively strong semantic density (SD+) as it has a relatively complex set of meanings associated with it. In contrast, the term *push* exhibits significantly weaker semantic density (SD−), as it does not resonate out to the same degree range of technical meaning in the field.[1]

Semantic gravity and semantic density can vary independently. On this basis, the *semantic plane* in Figure 7.1 (Maton 2016: 16) shows the possibilities for variation − with semantic gravity and semantic density as axes and each quadrant representing a different *semantic code*.

Rhizomatic codes (top-right quadrant) are meanings that maintain weaker semantic gravity (SG−) and stronger semantic density (SD+). In common-sense terms, this quadrant indicates meanings that are relatively 'generalized' and 'complex', as often exemplified by technical theory. In contrast, *prosaic codes* (bottom-left quadrant) are meanings with stronger semantic gravity (SG+) and weaker semantic density (SD−); meanings that are 'concrete' and 'simple', and often illustrated with common-sense everyday knowledge. *Worldly codes* (bottom-right quadrant) exhibit stronger semantic gravity and stronger semantic density; meanings that are both highly

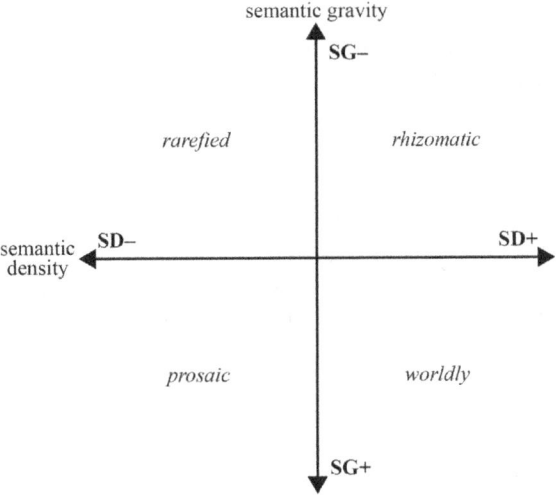

FIGURE 7.1 The semantic plane (Maton, 2016, p. 16)

context-dependent and very complex. This quadrant is often illustrated by knowledge in areas such as vocational education, and practically oriented fields and crafts, where projects are often embedded in a specific instance or case (such as building a particular bridge over a particular river) but involve a complex knowledge-base to achieve their goals. Finally, *rarefied codes* (top-left quadrant) are meanings with weaker semantic gravity and weaker semantic density; meanings that are relatively context independent but simple.

The *semantic plane* enables us to see the possible combinations and gradations available across the disciplinary map. Focusing on physics, it is tempting to immediately position it within the top-right quadrant as a *rhizomatic code* – as being both 'abstract' (weaker semantic gravity) and 'technical' (stronger semantic density). Indeed if we follow Biglan's (1973) characterization of physics as a relatively 'pure' discipline (i.e. non-applied) and a relatively 'hard' science, or Kolb's (1981) description of physics as both a relatively reflective (non-applied) but also a particularly 'abstract' discipline, then a classification of physics as a rhizomatic code makes sense. However, once we look at the actual practices of any field, it becomes clear very quickly that no discipline fits into a single box. There is immense variation across sub-disciplines (is it nuclear physics? classical mechanics? electromagnetism? astrophysics?); across year level (elementary? secondary? tertiary? which year in each?), across individual classes and across research, practice and education. LCT is invaluable here: the semantic codes are not boxes – the semantic plane is a topological space with indefinite variation *within* each code.

As we will see, one of the key shifts that occurs in physics texts from secondary education onwards is the ability to move between abstract theory and empirical instances while still using highly complex knowledge. Put another way, a key pattern is the varying of semantic gravity while maintaining relatively strong semantic density – shifts between rhizomatic codes and worldly codes. Similarly, when teaching new concepts, teachers and textbooks will regularly shift between 'everyday', 'concrete' knowledge of examples (prosaic codes) and the abstract theoretical knowledge typically considered the realm of physics (rhizomatic codes). In short, the key to doing physics is not just in being able to understand highly technical, abstract knowledge, but to vary the abstraction and technicality as required by the situation and to utilize the particular semiotic resources that organize this.

The following sections will step through the ways language, mathematics and image are used to organize both the technical meanings of physics and the shifts in semantic gravity and semantic density that allow students to bridge between 'theory' and the 'everyday', the 'empirical' and the 'abstract', and the 'concrete' and the 'complex'. First, it will focus on the developments in language that move knowledge from common-sense to technical understandings. Then it will turn to the move from language to mathematics in early secondary school and the reorganization of knowledge this entails. Finally, we will consider how the range of images used in physics complement the technical meanings and shifts in Semantics seen in mathematics. In sum, we will see that the three semiotic resources work together to organize the knowledge of physics in a way that allows students to build a scientific gaze.

Knowledge through language

Language and activity

An early stepping stone into physics is exemplified by a book for elementary (primary) school students, focusing on pushing and pulling (Riley 2001). This book gives a series of examples of pushing and pulling with very large photos of each action:

> *When do you push?*
> > *You push a pram*
> > *You push a swing*

> *When do you pull?*
> > *You pull a brush through your hair*
> > *You pull a book from your bag.*

This very concrete set of examples illustrates a series of physical events. In terms of the SFL register variable field, these events construe *activities* (see Doran and Martin, chapter 5 this volume). Activities such as these offer the child a dynamic perspective on the world as happenings. To build a picture of the activities involving pushing and pulling, the text lists a long series of examples with large pictures of each activity:

> *A digger pushes rocks into a heap.*
> *A toy car needs a push to make it go.*

Michael pushes the keys to play a tune. [referring to playing the keyboard]
Alex pushes the ball when he kicks it.
...
An engine pulls a train.
A tractor pulls a trailer.
A dog pulls on its lead.
Tim pulls on a jumper.

From the perspective of more advanced physics, this listing of examples appears relatively simple. However, it performs a crucial early role in building an uncommonsense understanding of the world. By listing a series of other quite different activities (kicking, playing with a car, piling rocks in a heap, putting on a jumper, driving a train) as 'pushing' and 'pulling', the text emphasizes their similarity and categorizes them as instances of the more general activity of pushing and pulling. In addition, the book illustrates that a wide range of *items* may do the pushing or pulling or be pushed or pulled. This is an important move into scientific knowledge as it means first that all examples may be discussed along similar lines, and second that pushes and pulls can be found almost everywhere. The student is learning that this wide range of activities, in some sense, all do the same thing.

In terms of LCT Semantics, this weakens the semantic gravity of knowledge. The individual examples describe specific events that are relatively context-dependent. But as more examples are listed, the differences between them are generalized in a way that emphasizes their similarities as pushes and pulls. That is, the text weakens semantic gravity so that it can eventually discuss pushes and pulls as events of their own (discussed below). Similarly, in terms of semantic density, by listing this series of examples as *pushes and pulls*, it relates these otherwise distinct events together. This adds connections between their meanings, slightly adding to their complexity, and thus strengthening their semantic density. Although these are relatively small shifts in semantic gravity and semantic density in comparison to higher level physics, they are important first steps in learning this knowledge which will repeated regularly as students move through school.

Looking further into the text, we see it also illustrates that pushing and pulling has effects on the world ('^' indicates that one activity follows or is entailed by the other):

You push on clay to squash it flat.

You push on clay
^ (in order to)
squash it flat
...
You pull an elastic band to stretch it.

You pull an elastic band
^ (in order to)
stretch it

...
> *Paul pushes the pedals on his bicycle. The wheels turn round.*

> *Paul pushes the pedals on his bicycle*
> ∧
> *The wheels turn round.*

Here, the text presents *squashing, stretching* and *turning* as resulting from the pushing or pulling in the previous clause. Series of events such as these where one activity implicates or results from another are a key feature of scientific discourse (Wignell *et al.* 1993). From the perspective of field, they show chains of activity that are implicated by one another.[2] Although at this stage each implication relation includes only two activities, later on in schooling these series become long and intricate, involving a range of possible dimensions of conditionality and causation (Rose 1998) and underpinning key genres in science such as explanations (Unsworth 1997, Martin and Rose 2008). Interpreting this from the perspective of LCT Semantics once more, these implication relations work to connect activities together, further strengthening their semantic density.

Activities and items

By generalizing a range of common-sense activities as 'pushing' and 'pulling', and linking them up with other activities to form series of implication activities, this primary school book has already taken some key steps in the development of an uncommon-sense scientific field. But the book does not stop here. After introducing a number of examples, the text nominalizes the activities of pushing and pulling as *pushes* and *pulls*, through a resource known as 'grammatical metaphor' (Halliday 1998):

> *A toy car needs <u>a push</u> to make it go*
> ...
> *What <u>pushes</u> and <u>pulls</u> can you find as you play?*
> ...
> *Can you think of three more <u>pulls</u>?*
> ...
> *Emma gives a weak <u>push</u> to her car.*
> *Ben gives a strong <u>push</u> to his car.*

In doing so, the text reconstrues the activity of *pushing* and *pulling* as an item *push* and *pull;* it turns an 'event' into a 'thing'. This is arguably one of the most significant moves in the transition to uncommon-sense knowledge (Halliday and Martin 1993, Halliday 1998), as it enables much greater possibilities for meaning than if it were dealing with *just* an activity, or *just* an item.

In this first instance, itemizing these activities enables them to 'do' other activities themselves. For example:

> *A push can squash something*
> …
> *A pull can stretch something*

Here, the push and pull have been abstracted from any particular thing doing the pushing and pulling. This establishes a more generalized series of activities whereby one activity (the *push* or *pull*) leads to another (*squashing* or *stretching*). This further weakens semantic gravity, as the pushes and pulls are no longer tied to any particular instances:

> *Push*
> ∧
> *squash something*

and

> *Pull*
> ∧
> *stretch something*

This text does not take the next step of also generalizing the squashing and stretching and putting the whole series of implication in one clause, such as in *squashes are caused by pushes* (a specific type of grammatical metaphor known as a logical metaphor; Hao 2018, Halliday 1998). But through this initial grammatical metaphor, the book establishes the basic building blocks that students need for a scientific construal of experience.

Second, by itemizing *pushing* and *pulling* as *push* and *pull*, the text enables them to enter into relations normally reserved for items. This is most clearly seen towards the end of the book, when it specifies that:

> *A push is a force*
> …
> *A pull is a force*

Here, the pushes and pulls are classified as types of *force*. In doing so, the book establishes a *classification taxonomy*. As a wide range of previous events have already been described as pushes and pulls, this means they are all in turn classified as types of *force*. The term *force* then resonates out to a wide range of common-sense experiences into a single technical term (Wignell *et al.* 1993). Again in terms of LCT Semantics, this strengthens its semantic density, establishing a relatively large series of interconnections emanating from the technical term *force* (see Maton and Doran 2017b). As students move through later years of school, *force* will become increasingly central to the field of physics as it distils more and more meaning.

Language and properties

There is one final step this text takes in construing its scientific view of the world. This can be seen in the following example where the text notes that pushes and pulls can be stronger or weaker:

> Emma gives *a weak push* to her car.
> Ben gives *a strong push* to his car.
> Ben's car goes further than Emma's car.

In the terms of Doran and Martin (Chapter 5 of this volume) model of field, *weak* and *strong* are *properties* of the pushes. Properties are meanings attached to an item or an activity that can be graded as more or less. As this instance shows, by attributing the property of strength to the activities, the two pushes can be ordered as more or less strong (or *strong* and *weak*) – what is referred to as an *array* of strength (Doran 2018). Importantly, this array of strengths can also have effects. In this case, the differences in strength lead to a difference in the distance each of the cars go. One array, the strength of the push, leads to another array, the distance of the movement, shown by *Ben's car goes further than Emma's car.*

This potential for giving items and activities a property, ordering these properties into arrays, and setting up chains of dependencies and causation between different arrays, opens a further avenue for building connections between meanings and strengthening semantic density.

Building the scientific gaze

Although at first glance the meanings being made at this level may appear relatively simple, this text has introduced each of the main basic building blocks that students will need throughout their scientific study. From the perspective of field, the text has explored a *dynamic* perspective on phenomena as activities of pushing and pulling, complemented with a *static* perspective of pushes and pulls as items, and added gradable properties of strength to the pushes and pulls. In addition, the text has introduced small sets of relations between each of these meanings. It has brought together activities into relations of *implication*, where one activity entails another (such as *pushing* leading to *squashing*); it has brought together items into a small *classification taxonomy* whereby one set of items are positioned as types of another (i.e. *pushes* and *pulls* as types of *force*); and it has ordered sets of properties in relation to each other along an *array* (where *pushes* and *pulls* are ordered as more or less *strong*). Table 7.1 synthesizes the field-specific meanings made through language to this point.

These elements of field underpin the content knowledge developed in science from primary school onward. As students move through schooling, implication series, uncommon-sense taxonomies and arrays will be extended and integrated; and as we will see in the following sections, they will be greatly elaborated through

TABLE 7.1 Elements of field in a primary school text

Elements of field		Examples
activity	single activity	*An engine pulls a train.*
	momented as an implication activity	*You push on clay to squash it flat.*
item	single item	*What <u>pushes</u> can you find as you play?*
	items in a (classification) taxonomy	*A push is a force.*
property	single property	*Emma gives a <u>weak</u> push to her car.*
	properties ordered into an array	*Ben's car goes <u>further</u> than Emma's car.*

the use of images and mathematics. However, the introduction of these meanings does more than simply lay out the specific types of technical meaning needed in science. They also provide early steps in shifts that underpin the scientific gaze. In particular, they show a pathway through which students can move from relatively everyday concrete meanings to more uncommon-sense meanings. More technically in terms of LCT, they model a pathway from relatively weak semantic density to slightly stronger semantic density by raising the complexity of meanings, and from relatively strong semantic gravity to slightly weaker semantic gravity by lessening the context-dependence of meanings.

In terms of semantic density, we have seen that the text establishes relations between examples first by likening them together as examples of pushing and pulling, and second by connecting them into classification taxonomies, arrays and relations of implication. This takes relatively distinct common-sense meanings and begins to establish more elaborated networks of meaning. The final step in strengthening semantic density in this text comes through the introduction of the term *force*. By classifying pushes and pulls as forces, the text condenses all of the meanings established through the book into a single term. This positions it as a key instance of technicality in the field of physics (Halliday and Martin 1993), which exhibits relatively strong semantic density (Maton and Doran 2017b). Although the field of physics at this stage is not particularly elaborated, the term *force* is the first to position the entire discussion within the technical domain of physics.

At the same time, the text weakens the semantic gravity of its knowledge. It generalizes a series of distinct examples as the 'same' events – pushing and pulling – thereby moving them further away from any particular instance. It then reconstrues these events of pushing and pulling as items: *pushes* and *pulls*. This has the effect of extricating *pushes* and *pulls* from any particular things doing the pushing and pulling, and ensures they can be discussed as items in themselves. Extending this, by technicalizing *pushes* and *pulls* as *forces*, the text further lessens its context-dependence, weakening semantic gravity. As Doran and Maton (2021) argue, positioning meanings within more elaborated networks of meaning within a particular field tends to stabilize them. It makes them more dependent on other meanings in the field than on any particular context in which it is used, and in doing so weakens its semantic gravity.

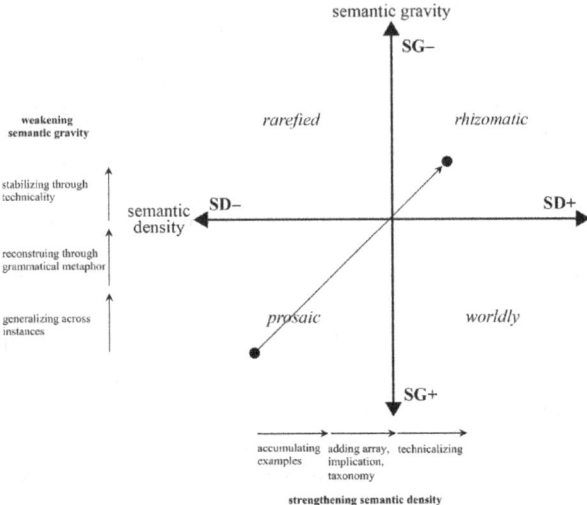

FIGURE 7.2 Movements in the semantic plane through the language of pushing and pulling

These movements in semantic gravity and semantic density are mapped in Figure 7.2. As the plane indicates, a number of the linguistic resources used have effects on both semantic gravity and semantic density. This means that for students reading this text, the scientific knowledge to be learned is two-fold. On the one hand, students are to learn the specific content meanings associated with pushing, pulling and force – considered here through the field relations organizing scientific meanings. On the other hand, underpinning this, students are to learn that scientific knowledge involves reconstruing everyday situations into less context-dependent and more complex meanings that can resonate out to a range of other situations. In terms of LCT, they are learning that the scientific gaze involves being able to weaken semantic gravity and strengthen semantic density across a range of situations.

Knowledge through mathematics

Language occurs across all levels of science. The patterns shown in primary school where common-sense meanings are reconceptualized into uncommon-sense, technical meanings recur innumerable times throughout schooling. With each passing year, the expanse of these meanings increases, with deeper and more integrated relations in field being built, significantly strengthening semantic density. All the while, everyday meanings with stronger semantic gravity are reconceptualized in terms of more generalized meanings with weaker semantic gravity. For students successful in accessing the knowledge of science, this pattern instils a gaze that values movements from prosaic code (SG+, SD−) conceptions of the world to rhizomatic code (SG−, SD+) reconstruals.

From the perspective of activity, the later years of schooling build much longer implication series in order to explain and predict phenomena. These are complemented by activities associated with experimental procedures and recounts, which involve expectancy series. In these activities, relations tend to unfold through time in terms of what is expected to happen, rather than via definite entailment (Hao 2020, Doran 2018). In terms of taxonomy, larger classification taxonomies are complemented by compositional taxonomies that construe parts to wholes (such as a nucleus to an atom).

In addition one of the major changes in the move from primary school to secondary school tends is often an increased emphasis on properties. As mentioned above, properties construe gradable phenomena – as in *the strength of a push*. These properties may be ordered into arrays – as in *one push is stronger than another*. Such arrays enable science to construe the world not just in terms of categorical distinctions, but also as involving infinitely variable gradations of more or less. This allows science to technicalize the 'fuzziness' of the physical world, where things are often distinguished by degree rather than by discrete oppositions.

From secondary school onwards, a regular resource for organizing properties is mathematical symbolism (Doran 2017). Each symbol in a mathematical statement realizes a property.[3] For example the equation $V = IR$ describes properties of electricity going through a conductor (known as Ohm's law). This equation involves symbols where there may be more or less voltage (V), more or less current (I) or more or less resistance (R). The equation thus construes variation that occurs in the world.

Importantly, placing these symbols into an equation indicates that these properties are not independent. For any particular instance, not every possible value is available for each symbol. Rather, the equation specifies a set of relationships between each of these symbols whereby if there is a change in the value of one symbol, it will likely affect all the others. For example, in $V = IR$, if I (the current) was to increase, then either V (voltage) would also need to increase, or R (resistance) would need to decrease, or both. Similarly, if V were to increase, then either I or R or both would also need to increase; and if R were to increase then either V would increase or I decrease or both (see Doran 2018: 88–96). As far as field is concerned, Doran and Martin (chapter 5, this volume) term these relations between properties *interdependency relations*. These interdependencies greatly expand the meaning potential of science. In the first instance, by involving properties, they allow scientific fields to construe degrees of gradation in ways that activity and taxonomy cannot. Secondly, by specifying definite interdependencies between these properties, they offer a means for describing the effects of a change in any particular property. Put another way, through the equation $V = IR$, physics is able to account for there being more or less voltage, more or less current and more or less resistance, and it is also able to precisely describe the effect of a change in voltage on the current or resistance of a conductor. By virtue of establishing interconnections between properties, mathematics also contributes to strengthening the semantic density of knowledge.

Mathematical equations can be organized into two key genres that deal with different aspects of scientific knowledge. These genres are known as *derivations*, which develop new relations between symbols, and *quantifications* which work to quantify symbols (Doran 2017, 2018). Derivations are concerned with deriving new equations from previously known ones. In New South Wales, Australia, they tend to appear in physics in the later years of secondary school. For example, in the following derivation from a secondary school student exam response, the student derives a new equation for W (*work*) (crudely, the energy associated with a force) in response to a question asking about the work needed to move a satellite from earth to an orbit altitude. Note that in the final equation, the W on the left side of the $=$ is elided by convention in mathematical texts (Doran 2018: 69–72):

$$work = \Delta GPE = GPE_f - GPE_i$$

$$\therefore W = -\frac{Gm_1m_2}{r_f} - \left(-\frac{Gm_1m_2}{r_i}\right)$$

$$= Gm_1m_2\left(\frac{1}{r_i} - \frac{1}{r_f}\right)$$

The student begins by stating an equation that is assumed at this level, namely that *work* is equal to the change in gravitational potential energy (ΔGPE, meaning roughly the change in energy associated with moving between different positions in the gravitational field of earth). From this initial relation, the student then inserts sets of other known relations – such as the fact that the change in gravitational potential energy (ΔGPE) is equal to the difference between the final gravitational potential energy (GPE_f) and the initial gravitational potential energy (GPE_i), and that gravitational potential energy itself is equal to $\frac{Gm_1m_2}{r}$ (glossed as the gravitational constant G, multiplied by the mass of the satellite (m_1), multiplied by the mass of the earth (m_2), all divided by the distance between the satellite and the centre of the earth (r)). In the final line, the student rearranges the equation to simplify it for calculations further along in the text. This derivation brings together a complex network of technical meanings and relates them precisely in terms of their interdependencies.

The specific relations in this derivation are not important for our discussion. What is important is that through the manipulation of mathematical symbolism the student is able to move from one set of relations, *work* $= \Delta GPE$, to a new set of relations $work = Gm_1m_2\left(\frac{1}{r_i} - \frac{1}{r_f}\right)$. In effect, the student has established new sets of interdependencies in the field; they have established relations not previously assumed (at this level, or at least logogenetically in this text), and in doing so, they have more tightly integrated the technical meanings of physics. In terms of LCT Semantics, derivations function to strengthen the semantic density of the discipline. They enable new relations to be established, which integrate an increasingly wide range of technical meanings. This means that any particular meaning specified in the derivation can now resonate out to all other meanings mentioned. But more than this, each term can be linked to other, unstated technical meanings by virtue of their relation any other symbol that is mentioned. This is a powerful means of strengthening semantic density, and one that appears to be increasingly relied upon as students move into higher levels of physics.

A little before derivations are introduced, students become familiar with another genre of mathematics known as a quantification. Where derivations build new, previously unknown relations, quantifications measure a particular property by numerically quantifying it. In terms of field, Doran and Martin (Chapter 5 of this volume) refer to the numerical measurement of properties as *gauging*. An example of this can be seen in the student exam response immediately following the derivation discussed above. Drawing on the final result of the derivation, the student inserts a set of numbers given in the question to calculate the work done in a particular situation:

$$[W] = Gm_1m_2\left(\frac{1}{r_i} - \frac{1}{r_f}\right)$$

$$= 6.67 \times 10^{-11} \times 1428.57 \times 5.97 \times 10^{24}\left(\frac{1}{6\,380\,000} - \frac{1}{6380000 + 355000}\right)$$

$$\doteqdot 4699722327$$

$$\doteqdot 4.70 \times 10^9 \text{ J} \left(3 \text{ sig fig}\right)$$

Here, through numerous steps (detailed in Doran 2018), the student concludes that the work done is approximately 4.70×10^9 Joules (~ 4.7 billion Joules). What is significant about this text is that the quantification has enabled the student to relatively precisely measure a specific instance of a property. It has, in other words, taken the relatively generalized relations given in the symbolic equations and applied them to a specific situation. The quantification genre has moved the text from the relatively weak semantic gravity descriptions of general relations between properties in physics, to relatively strong semantic gravity measurements of an individual property tied to a very specific situation.

The two mathematical genres, derivation and quantification, thus enable students to strengthen semantic density and strengthen semantic density – to build theory and to link this with the empirical world. Symbolic equations such as $V = IR$ are not tied to any particular instance. Rather, they describe abstracted relations that encapsulate an innumerable number of instances. In this sense, they maintain relatively weak semantic gravity. At the same time, they realize a definite set of technical relations between highly technical symbols and so involve relatively strong semantic density. Symbolic equations can thus be positioned in the rhizomatic code (SG–, SD+). Derivations progressively pull together more and more relations between an ever-widening number of technical symbols, but they do so in a way that is not heavily tied to any particular context or situation – they remain relatively generalized. In this sense, derivations strengthen semantic density of the field, while having little effect on semantic gravity.[4] Derivations can therefore be described as strengthening semantic density *within* the rhizomatic code (moving rightwards on the semantic plane). On the other hand, quantifications move texts from relatively weak semantic gravity relations that describe a wide range of phenomena to measurements tied specifically to a single instance. This means they significantly strengthen semantic gravity of the text. At the same time, they do this without decreasing the complexity of this knowledge. Moving from $[W] = Gm_1m_2\left(\dfrac{1}{r_i} - \dfrac{1}{r_f}\right)$ to $W \doteq 4.70 \times 10^9 \text{ J} \left(3 \text{ sig fig}\right)$ does not move us into the realm of everyday, common-sense meanings. The meanings maintain relatively strong semantic density by virtue of the fact that the meanings being quantified (i.e. the units of measurement and the organization of this measurement) are all highly technical and resonate out to the much wider disciplinary knowledge of physics. But they do so while significantly strengthening semantic gravity. In this sense, quantifications enable mathematics to move from *rhizomatic codes* into *worldly codes* – to move from theoretical to empirical (rather than from theory to the 'everyday'). This shows that the description of physics as a relatively 'abstract' discipline (e.g. Kolb 1981) misses a key component of its knowledge, namely that it enables increasing complexity of knowledge while shifting from 'abstract' theory to grounded empirical measurements. These movements in semantic gravity and semantic density through mathematics are shown in Figure 7.3.

For students learning this knowledge, the use of mathematics complements the trained gaze developed in language. In addition to the movements shown in language from a prosaic code (SG+, SD−) to a rhizomatic code (SG−, SD+), the discipline emphasizes the importance of both continually strengthening semantic density *within* the rhizomatic code and being able to reach down to the empirical

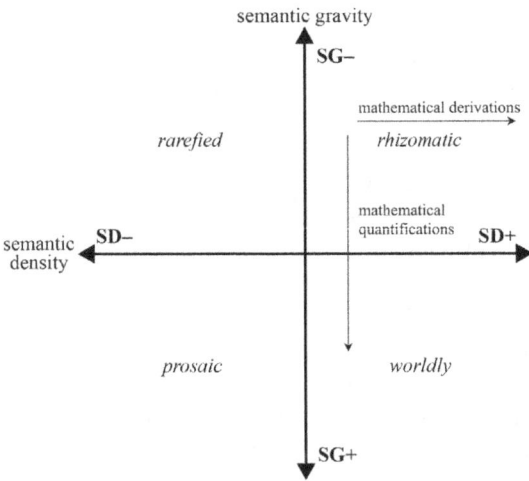

FIGURE 7.3 Movements in the semantic plane through mathematics

(worldly code or SG+, SD+), without moving into the common-sense. This has two effects. First, it emphasizes regular movement: the scientific gaze it is exhibiting is not one of static technicality or abstraction but of constant movement between the everyday, the theoretical and the empirical. Second, it emphasizes the utility of theoretical knowledge in terms of its ability to reach down and make precise predications about and descriptions of the empirical world. Such movements are vital for conceptualizing new situations from a scientific viewpoint.

This raises the question of how empirical meanings (SG+, SD+) can reach back to the theoretical meanings (SG−, SD+). We have seen that mathematics enables students to reach towards empirical description by quantifying individual properties but we have not yet seen how physics can use empirical measurements to change theory. In terms of semantic gravity, mathematics enables physics to strengthen semantic gravity (and to move from rhizomatic codes to worldly codes) but it offers no way at this stage to weaken semantic gravity again (to move back from worldly codes to rhizomatic codes). To deal with this issue, physics brings in images – specifically graphs.

Knowledge through image

Images occur throughout science (Parodi 2012, Lemke 1998). They enable a large number of meanings to be brought together and presented in a single, synoptic snapshot (Doran 2019, Martin *et al.* 2021). We can see this in the diagram in Figure 7.4, from the same student exam response as the mathematical examples above. In this question, the student was asked to *Draw a labelled diagram of the vacuum tube used by Thomson to calculate the q/m ratio of electrons.*

From the perspective of field, the diagram depicts a compositional taxonomy, in which each component is positioned as a part of the apparatus (the whole). However, the diagram does not just present the various parts without any sense of

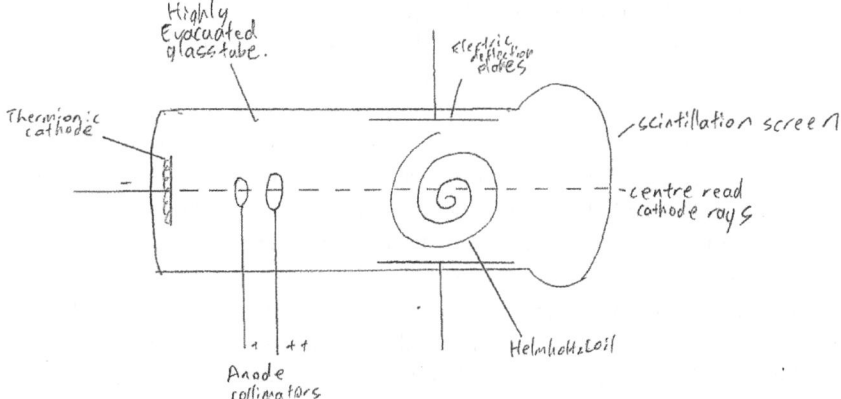

FIGURE 7.4 Diagram of an experimental apparatus

how they come together to make the whole, what Kress and van Leeuwen (2006) call an *unstructured analytical* image. Rather, the components are spatially arranged in relation to one another – known as a *spatially structured* analytical image. In terms of the field relations of Doran and Martin (chapter 5, this volume), this means each component of the image realizes two distinct meanings: its part within a compositional taxonomy and its spatial position in relation to all the other parts in a spatial array.[5] The image also depicts activity through the dotted line labelled *centre read cathode rays*. Although there is no arrow or other overt indicator of a vector, the linear arrangement of the terms *cathode* and *anode* – as technical meanings in this field – are enough for an informed reader to understand the movement of the cathode rays is from left to right.

This diagram thus integrates three distinct dimensions of field: a compositional taxonomy incorporating each part of the apparatus; a spatial array that arranges the parts of the apparatus in relation to one another; and an activity involving a cathode ray that moves through the apparatus. It is difficult for either language or mathematics alone to bring together this wide range of meaning in one snapshot. Indeed, this is one of the key affordances of images: they are able to pull together a large number of complementary meanings in a synoptic 'eyeful'. Diagrams in this sense are a key resource for packaging up multiple meanings to be read and viewed together. In terms of LCT, they offer the potential for significantly stronger semantic density in a way that complements that of language and mathematics.

As students move further into physics study, these diagrams are typically complemented by graphs. Graphs offer a relatively unique affordance in physics in comparison to the other resources we have seen so far and appear to be especially prevalent as students engage more deeply with experimental investigations. In terms of their knowledge-building potential, graphs complement mathematical quantifications by enabling a more flexible interaction between empirical measurements and more abstract theory. We can see this by exploring Figure 7.5. This figure displays a question from the same high-school student exam response as the diagram and the mathematical examples above. The prompt includes a table of measurements for Resistance and Temperature of a wire. The question asks the student to: (1) plot these values on the graph (shown by the 'x's on the graph); (2) draw a line of best fit (that generalizes across the 'x's, shown by the line labelled LOBF); and (3) use this line of best fit to estimate the resistance of the wire at 30°C (shown below the graph).

From the perspective of field, the graph is organized around two properties presented on the axes (Resistance and Temperature) which are given measured values in the table above the graph. The first task of the student is to take these measurements in the table and plot them on the graph (shown by the 'x's on the graph). From the perspective of LCT Semantics, these plotted measurements show relatively strong semantic gravity. Each point describes a very precise instance that cannot be generalized to any other. But, like the numerical measurements in quantifications, they also exhibit relatively strong semantic density as they signify the intersection of two technical properties and are measured with the particular units of these properties (Ohms for Resistance, Degrees Celsius for Temperature).

The electrical resistance, R, of a piece of wire was measured at different temperatures, T. Near room temperature, the resistance of the wire can be modelled by the equation $R = mT + b$.

Temperature (°C)	Resistance (ohms)
12.5	0.122
16.4	0.124
32.6	0.130
36.5	0.131

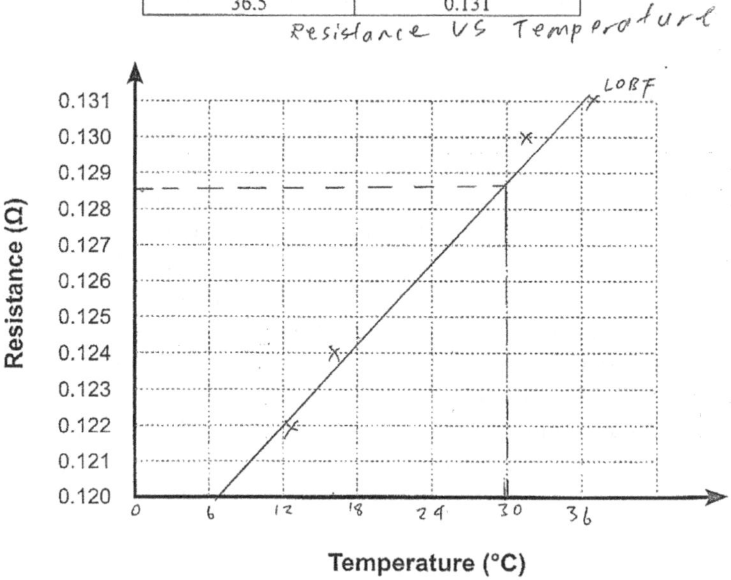

(a) Plot the data points on the graph provided. Draw a line of best fit on the graph and use it to estimate the electrical resistance of wire at 30° C.

using extrapolation $\div$ *0.128.5 Ω*
$\div$ *0.129 Ω (3 sig fig).*
10

FIGURE 7.5 Graph and questions in a senior secondary school examination

The next step for the student is to draw a 'line of best fit'. This involves generalizing across the plotted points to draw a line that 'best fits' each of the values. In this case, this is a straight line drawn from about six degrees Celsius on the temperature scale to the left, up to about 37 degrees in the top right. Like mathematical equations, this line realizes a general interdependency between properties. It shows that when resistance increases, so does temperature (and vice versa). Similarly, this line has significantly weaker semantic gravity than the individual plotted points. Rather than being tied to any actual instance, the line develops a more generalized description. Indeed, the line does not intersect precisely with any of the measured points. Drawing a line of best fit weakens semantic gravity such that the relations it describes are not dependent on any particular instance or context.

By shifting between measured points and generalized lines of best fit, graphs can be used to shift from stronger semantic gravity to weaker semantic gravity while maintain relatively strong semantic density. They thus complement mathematical quantifications by allowing for students to shift from the worldly code back into the rhizomatic code. They enable a shift from the 'empirical' back up to the 'theoretical'.

An important feature of images is that they do not insist on a definite reading path (Bateman 2014). In the case of graphs, this means students do not just have to move from measured points to generalized lines (from stronger to weaker semantic gravity). They can also use the generalized line to predict specific instances. Indeed, this is the third task the student is asked to do, when question (a) says to *estimate the electrical resistance of wire at 30° C.* To answer this question, the student would have to find where the line of best fit is at 30° C, and 'read off' the value of the Resistance at this point. In this example, the student has done this by drawing two dotted lines, one vertical line at 30° Celsius, and then one horizontal line at around .1285 Ohms Resistance. Below the graph, the student then rounds this up to 0.129 Ohms. In terms of semantic gravity, this displays a movement back from weaker semantic gravity to stronger semantic gravity. The student begins with the generalized description shown by the line of best fit and uses this to make a prediction of a specific measurement. Although this is an estimation, the prediction is strongly dependent on a context where the wire is 30°C, and so it displays relatively stronger semantic gravity.

This example illustrates how graphs enable movements back and forth between stronger semantic gravity and weaker semantic gravity. This affordance makes it a key tool for organizing the knowledge of physics (and many other disciplines). Figure 7.6 illustrates this movement. Where mathematical derivations enable movements within the rhizomatic code, and quantifications organize one-way trips from the rhizomatic code to the worldly code, graphs give opportunities for shifts back and forth – what Maton and Howard (2018) term 'return trips'.

Multimodal knowledge

We can now return to the original question of this chapter: what knowledge do students need to learn in order to be successful in science? In the first instance, they need to understand the technical content meanings of the discipline. From the perspective of field, this means marshalling various items, activities, properties, and their respective relations when necessary. As this chapter has shown, this necessarily implicates multiple semiotic resources. Mathematics is able to bring together rich interdependencies between properties and measure these quantitatively; graphs are able to establish multiple arrays upon which measurements can be ordered; and diagrams and language can integrate a wide range of different types of meaning,

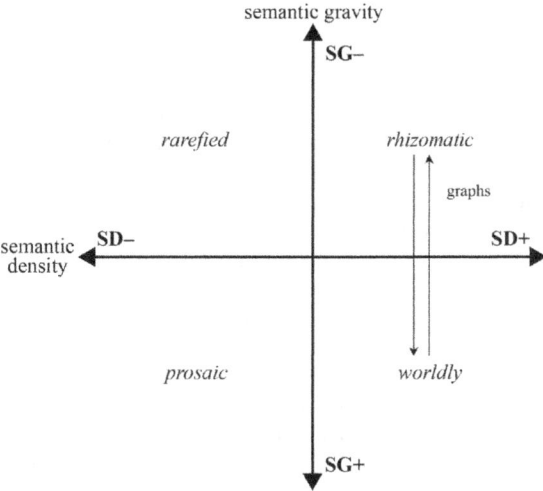

FIGURE 7.6 Movements in the semantic plane through graphs

including taxonomy, activity and property. The content meanings of science are organized through the multimodal resources of science.

In the second instance, students need to learn a particular way of seeing the world. This way of seeing the world does not simply involve technical discussions of abstract phenomena. More importantly, it involves moving between the 'common-sense' and the 'technical', the 'theoretical' and the 'empirical'. In terms of LCT Semantics, this means being able to move between a prosaic code (SG+, SD−) and a rhizomatic code (SG−, SD+), and between a rhizomatic code and a worldly code (SG+, SD+). Or, to put this another way, students need to be able to strengthen and weaken semantic gravity and semantic density independently and together. They need a way of seeing the world from various perspectives and shifting between these perspectives as needs arise. As mentioned above, what is at stake here is a trained scientific gaze. Being able to read and write particular the semiotic resources and field-specific meanings of science is not enough. Students need to be able to select and arrange these resources appropriately in potentially new scenarios. This means they need a way of understanding the organizing principles of the knowledge in any particular situation. For physics (and science more broadly), this necessarily involves using multiple semiotic resources. Language enables the common-sense to be repackaged as uncommon-sense (and vice versa); mathematics enables a steady increase in the complexity of meaning while also moving from the theoretical to the empirical; and images enable an integration of a wide range of meanings to increase complexity and movement back and forth between the empirical to the theoretical. In all, students must learn that scientific knowledge is multimodal.

Notes

1 The semantic density and semantic gravity explored in this chapter are those associated with 'content' meanings – technical procedures, logical reasonings, taxonomies, etc. This is known as *epistemic–semantic density* and *epistemic–semantic gravity* (Maton 2014, Maton and Doran 2017b). This contrasts with *axiological–semantic density* and *axiological–semantic gravity* that explore meanings associated with values, political judgements, morals, aesthetics, etc.

2 These were previously called implication *sequences* (e.g. Martin 1992). However, following Hao (2015, 2020) 'sequence' is reserved for relations between figures in discourse semantics, and 'implication' will be reserved for one type of relation between activities in a field. The relationship between a larger activity and the smaller activities that constitute it is known as *momenting* (discussed in Doran and Martin, Chapter 5 of this volume).

3 More precisely, each symbol realizes an *itemized property* – a property reconstrued as an item. This is because they may be graded (like properties), but they can also be classified (like items). For example, one may have more or less E (energy) (interpreted as a property), but one may also distinguish between different types of energy, such as initial energy, E_i, final energy E_f and average energy E_{av} (interpreted as items in a classification taxonomy). See Doran and Martin, chapter 5 of this volume for discussion.

4 It is possible that particular derivations may strengthen or weaken semantic gravity by either deriving a set of relations only applicable to more precise situations or conversely deriving more generalized relations to cover a wider set of phenomena. This would be a function of the particular derivation under study rather than all derivations in general.

5 More precisely, each element realizes a position along two spatial arrays, one vertical and the other horizontal.

References

Bateman, J. (2014) *Text and Image: A Critical Introduction to the Visual/Verbal Divide*, London: Routledge.

Biglan, A. (1973) 'The characteristics of subject matter in different academic areas', *Journal of Applied Psychology*, 57(3): 195–203.

Christie, F. and Martin, J. R. (1997) (eds) *Genre and Institutions: Social Processes in the Workplace and School,* London: Continuum.

Christie, F. (2002) *Classroom Discourse Analysis: A functional perspective*, London: Continuum.

Doran, Y. J. (2017) 'The role of mathematics in physics: Building knowledge and describing the empirical world' *Onomázein Special Issue*, March: 209–26.

Doran, Y. J. (2018) *The Discourse of Physics: Building Knowledge through Language, Mathematics and Image*, London: Routledge.

Doran, Y. J. (2019) 'Building knowledge through images in physics', *Visual Communication*, 18(2), 251–77.

Doran, Y. J. and Maton, K. (2021) '*Context-dependence: Epistemic-semantic gravity in English discourse*', *LCT Centre Occasional Paper 2*

Halliday, M. A. K. (1998) 'Things and relations: Regrammaticizing experience as technical knowledge', in J. R. Martin & R. Veel (eds) *Reading Science*, London: Routledge, 185–235.

Halliday, M. A. K. (2004) *Language of Science,* London: Continuum.

Halliday, M. A. K. and Martin, J. R. (1993) *Writing Science: Literacy and Discursive Power*, London: Falmer.

Hao, J. (2015) 'Construing biology: An ideational perspective', Unpublished PhD thesis, Department of Linguistics, University of Sydney.

Hao, J. (2018) 'Reconsidering "cause inside the clause" in scientific discourse – from a discourse semantic perspective in systemic functional linguistics', *Text and Talk*, 38(5): 525–50.

Hao, J. (2020) *Analysing Scientific Discourse from a Systemic Functional Linguistic Perspective*, London: Routledge.

Kress, G., Jewitt, C., Ogborn, J. and Charalampos, T. (2001) *Multimodal Teaching and Learning: The Rhetorics of the Science Classroom*, London: Continuum.

Kress, G. and van Leeuwen, T. (2006) *Reading Images: The Grammar of Visual Design*, London: Routledge.

Kolb, D. A. (1981) 'Learning styles and disciplinary differences', in A. Chickering (Ed.) *The Modern American College*, San Francisco: Jossey-Bass, 232–55.

Lemke, J. L. (1990) *Talking Science: Language, Learning and Values*, Norwood, NJ: Ablex.

Lemke, J. L. (1998) 'Multiplying meaning: Visual and verbal semiotics in scientific text', in J. R. Martin and R. Veel (eds) *Reading Science*, London: Routledge, 87–113.

Lemke, J. L. (2003) 'Mathematics in the middle: Measure, picture, gesture, sign, and word', in M. Anderson, A. Saenz-Ludlow, S. Zellweger and V. Cifarelli (eds) *Educational Perspectives on Mathematics as Semiosis*, Ottawa: Legas, 215–34.

Martin, J. R. (1985) *Factual Writing: Exploring and Challenging Social Reality*, Geelong: Deakin University Press.

Martin, J. R. (1992) *English Text: System and Structure*, Amsterdam: John Benjamins

Martin, J. R. and Doran, Y. J. (2015) (eds) *Language in Education*, London: Routledge.

Martin, J. R., Maton, K. and Doran, Y. J. (2020) (eds) *Accessing Academic Discourse: Systemic Functional Linguistics and Legitimation Code Theory*. London: Routledge.

Martin, J. R. and Rose, D. (2008) *Genre Relations: Mapping Culture*, London: Equinox.

Martin, J. R. and Veel, R. (1998) (eds) *Reading Science: Critical and Functional Perspectives on Discourses of Science*, London: Routledge.

Martin, J. R., Unsworth, L. and Rose, D. (2021) 'Condensing meaning: Imagic aggregations in secondary school science', in G. Parodi (ed.) *Multimodality*, London: Bloomsbury.

Maton, K. (2014) *Knowledge and Knowers: Towards a realist sociology of education*, London: Routledge.

Maton, K. (2016) 'Legitimation Code Theory: Building knowledge about knowledge-building', in K. Maton, S. Hood and S. Shay (eds) *Knowledge-Building*, London: Routledge, 1–23.

Maton, K. (2020) 'Semantic waves: Context, complexity and academic discourse', in Martin, J. R., Maton, K. and Doran, Y. J. (eds) *Accessing Academic Discourse,* London, Routledge, 59–85.

Maton, K. and Doran, Y. J. (2017a) SFL and code theory, in Bartlett, T. and O'Grady, G. (eds) *Routledge Systemic Functional Linguistic Handbook*, London, Routledge, 605–18.

Maton, K. and Doran, Y. J. (2017b) 'Semantic density: A translation device for revealing complexity of knowledge practices in discourse, part 1 – wording', *Onomázein Special Issue*, March: 46–76.

Maton, K. and Howard, S. K. (2018) '*Taking autonomy tours: A key to integrative knowledge-building', LCT Centre Occasional Paper 1*: 1–35.

O'Halloran, K. L. (2005) *Mathematical Discourse: Language, Symbolism and Visual Images*, London: Continuum.

Parodi, G. (2012) 'University genres and multisemiotic forms of knowledge: Accessing specialized knowledge through disciplinarity', *Fórum Linguístico*, 9(4): 259–82.

Riley, P. (2001) *Ways into Science: Push and Pull*, London-Sydney: Franklin Watts.

Rose, D. (1998) 'Science discourse and industrial hierarchy', in J. R. Martin and R. Veel (eds) *Reading Science,* London: Routledge, 236–65.

Rose, D. (2004) 'Sequencing and pacing of the hidden curriculum: How indigenous children are left out of the chain', in J. Muller, A. Morais and B. Davies (eds) *Reading Bernstein, Researching Bernstein*, London: Routledge, 91–107.

Rose, D. (2014) 'Analysing pedagogic discourse: An approach from genre and register' *Functional Linguistics*, 2(11): 1–32.

Rose, D. (2020) 'Building a pedagogic metalanguage I: Curriculum genres', in J. R. Martin, K. Maton and Y. J. Doran (eds) *Accessing Academic Discourse,* London: Routledge. 236–67.

Rose, D. and Martin, J. R. (2012) *Learning to Write, Reading to Learn: Genre, Knowledge and Pedagogy in the Sydney School*, London: Equinox.

Rose, D., McInnes, D. and Korner, H. (1992) *Scientific Literacy (Literacy in Industry Research Report: Stage 1)*, Sydney: Metropolitan East Disadvantaged Schools Program.

Unsworth, L. (1997) '"Sound" explanations in school science: A functional linguistic perspective on effective apprenticing texts', *Linguistics and Education,* 9(2): 199–226.

Unsworth, L. (2001a) 'Evaluating the language of different types of explanations in junior high school science texts', *International Journal of Science Education,* 23(6): 585–609.

Unsworth, L. (2001b) *Teaching multiliteracies across the curriculum: Changing contexts of text and image in classroom practice*, Buckingham: Open University Press.

Wignell, P., Martin, J. R. and Eggins, S. (1993) 'The discourse of geography: Ordering and explaining the experiential world', in M. A. K. Halliday and J. R. Martin (eds) *Writing Science*, London: Falmer, 151–83.

PART III
Pedagogy in science education

8

WIDENING ACCESS IN SCIENCE

Developing both knowledge and knowers

Karen Ellery

Introduction

Historically universities served the interests of an élite few from the middle and upper classes of society. Thanks to the global economic and social justice drivers, massification of universities in the last century has resulted in major transformations across the higher education sector. A significant feature of this transformation has been greater diversity of the student body in terms of social and educational backgrounds. However, for many universities the fundamental 'culture' of teaching, learning and research has remained remarkably constant (Longden 2006), resulting in an inequitable system that favours certain social groups over others (Archer *et al.* 2003, Arum *et al.* 2012). Physical access without concomitant success is meaningless, and the emotional (Pym and Kapp 2011) and the financial toll (Council on Higher Education Report 2013) on individuals and families makes equity of outcomes an urgent moral imperative. The broader-scale implications of perpetuating a system with a grossly unequal distribution of power and access to educational, social and economic resources in any society add impetus to this imperative.

This chapter is particularly interested in access for success in the sciences. New scientific knowledge is constantly being developed and a typical pedagogic response in the higher education context has been to maintain the focus on knowledge but at the same time increase the volume and pacing of curriculum material. Muller (2015: 410) suggests this favours students from privileged educational backgrounds as they are 'better equipped by virtue of being educated in cognitively rich environments by better qualified teachers to respond to the increased volume of novel material'. There are, however, increasing calls from knowledge-based fields, such as the sciences and engineering, to consider ontological aspects of student 'becoming' and 'being' and agency in promoting learning (Barnett 2007, Dall'Alba and Barnacle 2007). In an in-depth study of student learning in engineering, Case (2013) makes an argument for student agency as a central part of engagement with, and learning of, requisite knowledge. As such, a student's social, cultural and educational background would strongly influence her/his engagement with a curriculum, and Case suggests that while engineering (or science) curricula must focus on the knowledge, there needs to be better accommodation of student identity development and agency. In short, she calls for more focus on both the knowledge and the knower.

The main question that this chapter addresses is how access for success can be enabled or constrained in a science curriculum.[1] To explore these issues, access for success is encapsulated in the term 'epistemological access' which Morrow (2009: 77) defines as learning 'how to become a participant in an academic practice' and that this requires learning 'the intrinsic disciplines and constitutive standards of the practice'. By invoking 'disciplines' and 'standards' of practice he is suggesting an underpinning framework, but his work does not provide the necessary analytical tools to unpack them. For this I turn to Legitimation Code Theory, which is a conceptual framework that provides a means for identifying the organizing principles that constitute curriculum practices. This chapter draws on a single case study of a higher education science access course that has been designed to accommodate students from traditionally less privileged education backgrounds. The findings suggest that the curriculum of the course involves two main bases for achievement, one which emphasizes the possession of knowledge, skills or procedures and another which emphasizes the need to be a particular kind of knower. Moreover, the course requires students to be the right kind of knower in order to then access the right kinds of knowledge. These findings are used to develop a conceptual model of epistemological access that could inform curriculum transformation processes in the sciences in this age of diversity and difference in higher education.

Conceptual and analytical framework

Legitimation Code Theory (LCT) is a conceptual framework that allows 'knowledge practices to be seen, their organizing principles to be conceptualized, and their effects to be explored' (Maton 2014: 3). The notion of legitimation is central to LCT: when actors engage in a practice, they are making a claim of legitimacy for the basis of that practice. What actors say or do can therefore be described as 'languages of legitimation', and the principles that underpin those practices are conceptualized as 'legitimation codes' (Maton 2016: 10). These codes represent the means by which success can be achieved in the practice. In this chapter, the knowledge practice that is the focus of our attention is the curriculum of an access course, and the legitimation codes that will help explore the role of constructions of knowledge and knowers in this curriculum are from the LCT dimension of Specialization (Maton 2014).

Specialization explores a set of organizing principles underlying practices as 'specialization codes'. These concepts begin from the simple point that practices are 'oriented towards something and by someone' (Maton 2016: 12), which highlights an analytic distinction: *epistemic relations* (ER) between practices and their object (that part of the world towards which they are oriented); and *social relations* (SR) between practices and their subject, author or actor (who is enacting the practices). The relative strengths of epistemic relations and social relations can vary from stronger (+) to weaker (–) along a continuum of strengths. Both refer to the basis of knowledge practices. For example, if your knowledge of environmental conditions (soil-type, nutrients, temperature, light conditions) promoting optimal growth of a particular plant was based on scientific knowledge, skills and procedures, it would represent stronger epistemic relations (ER+). In contrast, understanding that the plant grows best against the sunny wall of your house because it did when you tried it would represent relatively weaker epistemic relations (ER–): there is relatively less emphasis here on specialist knowledge, skills or procedures. Conversely, that experiential basis for deciding where to place your plant represents stronger social relations (SR+): it emphasizes your own experiences and opinions. The specialist scientific knowledge of optimal growth conditions downplays such personal opinions and experiences, representing weaker social relations (SR–). In identifying the relative strengths of epistemic relations and social relations, the *basis* upon which success in the practice is achieved is relevant as opposed to the sometimes more obvious *focus*. For example, student activities in a plant growth experiment may focus on recording plant size, leaf colour and flower production, but the basis for achievement in the final report may instead be the correct presentation of data.

By plotting the relative strengths of epistemic relations and social relations on a two-dimensional plane as in Figure 8.1, four principal codes are identified: knowledge codes, élite codes, knower codes and relativist codes (Maton 2014). When a practice has relatively strong epistemic relations and relatively weak social relations,

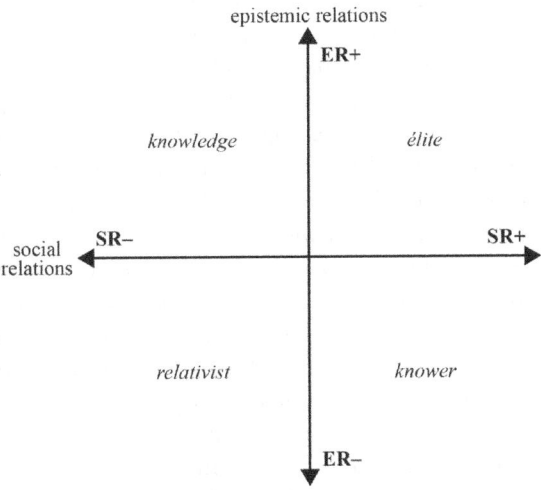

FIGURE 8.1 The specialization plane (Maton 2016: 12)

it legitimates a *knowledge code* (ER+, SR−). That is, the basis for specialization and legitimacy in the practice depends on possession of determinate and specialized knowledge and practices (ER+), and the attributes of knowers, such as their disposition and 'gaze', are downplayed (SR−). Science and science disciplines, such as physics and zoology, are typically considered to be dominated by knowledge codes. Educational practices characterized by a *knower code* (ER−, SR+) have relatively weak epistemic relations, but relatively strong social relations. Here, specialized knowledge is downplayed (ER−) and legitimacy is based on the disposition or 'gaze' of the knower (SR+). The humanities and many of their disciplines, such as sociology and anthropology, are often considered to be dominated by knower codes. Educational practices characterized by an *élite code* (ER+, SR+) have relatively stronger epistemic relations and relatively stronger social relations. In other words, both specialist knowledge and knower dispositions are equally valued. Professional levels of classical music, for example, require both depth of knowledge and personal talent – an élite code (Lamont and Maton 2008). Finally, practices characterized by a *relativist code* (ER−, SR−) have relatively weak epistemic relations and relatively weak social relations. Legitimacy here requires neither specialist knowledge nor knower dispositions. This is unlikely to be illustrated by subject areas in a specialized context such as education but can be found during periods of unfocused classroom activity or potentially during periods of 'brainstorming' when all ideas are welcomed.

It is worth emphasizing that these examples are simply illustrative: most sets of practices are likely to involve more than one code, and the particular code characterizing science, for example, depends on the context. While science is typically considered to be dominated by knowledge codes, in a specific classroom it may involve a range of codes, of which the dominant code may not be a knowledge code. Maton and Howard (2016), for example, found that a significant proportion of students' perceptions of secondary school science in New South Wales, Australia, exhibited a relativist code. The point is that these concepts enable us to explore these issues, rather than allocate subjects in advance to particular labels.

The specialization code exhibited by a curriculum may be expressed in explicit ways and therefore be obvious to participants, but may also be tacit, especially when underpinned by poorly articulated norms and values. Building on Bernstein's distinction between 'recognition' and 'realization' (2000), LCT suggests that for a student to be successful in a context they need to not only recognize its dominant code but must also be able to realize or enact practices that match that code. LCT therefore not only allows for identification and explicit articulation of the codes, but also opens up opportunity to see possible points of conflict. For example, if a science curriculum requires the use of empirical data to develop an argument, a knowledge code is being legitimated. However, if instead personal opinion is provided, the student is invoking a knower code. The resultant 'code clash' (Lamont and Maton 2008) can help explain poor student achievement. This study identifies specialization codes of a science course curriculum and invokes codes clashes to explain difficulties students have in their studies.

Context of study

Students' backgrounds

The study was located within a South African higher education science access programme. In order to enter any of the mainstream disciplines in science, students in the programme are required to pass three year-long foundation courses: Mathematical Foundations, Computer Skills for Science, and Introduction to Science Concepts and Methods (henceforth 'ISCM'). ISCM is the focus of this study. Because of a range of entrance selection criteria, students in the programme have achieved only slightly lower school marks than those entering mainstream directly, seldom have their home language as English, regularly require government funding to help pay for their tertiary studies, and attended poorly resourced and performing schools. They enter university with a range of literacy, numeracy and learning practices that are very different from the open-minded, analytical, creative, critical and independent approaches required when working with the high levels of conceptual knowledge in the sciences in higher education.

Study site

The purpose of ISCM, a multidisciplinary, integrated science foundation course, is to enable epistemological access to mainstream science by introducing students to scientific concepts, methods and literacies within the context of specific science disciplines, as well as through developing academic learning competencies and practices (Science Extended Studies Programme Review Report 2011). Currently ISCM draws on physics, chemistry, life sciences (human kinetics and ergonomics) and earth sciences (geology) for disciplinary input by mainstream staff from each of their respective disciplines. The focus in the two disciplinary lectures and one practical session per week is therefore on developing understanding of disciplinary concepts. Two permanently employed access programme lecturers facilitate the rest of the ten weekly contact periods, in what are termed 'literacies' pedagogic interactions.[2] While there is much overlap in the work of these two staff members, the 'language-related literacies' facilitator focuses mainly on reading and writing in the sciences. In contrast, the scientific-related literacies facilitator (myself) focuses on building foundational science concepts (such as spatial and temporal scales, hierarchies and connections, diversity) and procedures (such as working accurately and precisely, solving problems, working with empirical data) and considering how scientific knowledge is constructed. At least two of the weekly pedagogic interactions, particularly in the first semester, focus primarily on student learning in which strategies and approaches to learning are explicitly voiced, and appropriate practices related to student learning are modelled. As a coordinator of the course, I attend and observe or participate in all disciplinary interactions and some of the literacies interaction taught by my colleagues.

ISCM therefore has two main themes in its teaching: scientific and disciplinary concepts and methods; and student learning competencies. While these themes are often addressed separately in particular pedagogic interactions, the curriculum has been developed as a coherent whole and there is much integration across themes.

Methodology

The overall approach in the study was a single, in-depth case study. In order to surface the specialization codes that underpin the ISCM curriculum, data for epistemic relations and social relations analysis were obtained from course documents (handouts, resource materials, assessment tasks, student answers in assessment tasks), informal observations of pedagogic interactions, and interviews with four disciplinary and one literacies lecturer in which their perceptions of the purpose of ISCM were discussed. In order to bridge the gap between the empirical data and the abstract specialization concepts, a 'translation device' (Maton and Chen 2016) was developed through iterative movement between theory and data. In the original analysis, the strengths of epistemic relations and of social relations were examined independently of each other (Ellery 2017a, 2018) and thereafter specialization codes were identified (Ellery 2017b). What follows is the final code categorizations. In addition, in order to understand student response to codes, their answers in assessment tasks were analyzed and follow-up interviews with 17 volunteers of 47 students probed their approach to their studies (Ellery 2016, 2017b).

Specialization codes of ISCM

Because ISCM has an explicit dual focus of introducing students to disciplinary and scientific concepts, methods and literacies, as well as developing academic learning competencies and practices, two distinct codes were recognized: a *science-related knowledge code* and an *academic practices-related knower code*. These are now discussed in turn.

Science-related knowledge code

Since science and science disciplines are often associated with knowledge codes, it is not unexpected that such a code is highly visible in the ISCM curriculum. However, because the nature of the work done by the disciplinary and literacies lecturers in the course is so different, and because the strengths of epistemic relations for these two sets of work also vary, they are considered separately below.

Disciplinary epistemic relations and social relations

Evidence for the disciplinary component of the science-related knowledge code relates primarily to teaching done by mainstream disciplinary lecturers. Their focus on specialized and distinct concepts and procedures linked to their own disciplines, with little overlap between the work of the different lecturers, indicates relatively very strong epistemic relations (ER++). Table 8.1 outlines examples of disciplinary concepts identified in ISCM course documents, such as the geological processes of river formation, the concept of a mole in chemistry, the process of radioactive decay in physics, and factors that affect range of joint motion in human kinetics and ergonomics. Linking with these respective examples of disciplinary concepts, students are required to engage with specialized procedures such as plotting a river course on a topographic map, titrating a solution to determine molar concentration, calculating a radioactive decay constant and measuring range of joint motion using a goniometer. Assessment tasks that draw on these concepts and procedures consist of both low-order questions such as 'State the three main processes involved in river formation' and high-order questions such as 'Using a well-annotated diagram, explain the process of subduction in plate tectonics'.

While epistemic relations are strongly legitimated, social relations in the disciplinary component of the science-related knowledge code are not. Student dispositions, behavioural attributes and opinions are downplayed, representing weaker social relations (SR−). Typical examples, as indicated in Table 8.1, are where student opinion is sought on the effect of genetically modified organisms on the environment, or the use of nuclear energy as a form of energy generation, but students are judged on the coherence of their scientific argument, not on their opinion. Another example relates to the geological field trip in which the stated primary purpose is to enable 'careful, rigorous and systematic observation', which could represent stronger social relations linked to a particular way of working, but marks are allocated instead for correct identification of certain features and correct use of geological descriptors of rock characteristics, representing weaker social relations.

Scientific literacies epistemic relations and social relations

Evidence for the scientific literacies component of the science-related knowledge code relates mainly to work done by the two literacies lecturers. Since this work is underpinned by specialized science concepts, it represents stronger epistemic relations, but because many of the principles in this category can apply across a number of science disciplines, they are not as strong as the disciplinary-based category. Table 8.1 presents some examples of scientific literacies concepts that are visible in the ISCM course documentation. These include developing an understanding of how scientific knowledge is generated (for example through careful measurement and observation, inductively through looking for patterns in nature, deductively through generating hypotheses and conducting controlled experiments, and making predictions) as well as the basis upon which knowledge claims are made (for example, through using empirical data, recognizing the tentative nature of science, recognizing the idealization of many science laws, and knowing that understanding is often based on models of reality). Specialized procedures are closely linked to these concepts and relate to, among other things, designing and conducting experiments, developing coherent arguments, working with data (collecting, analysis, interpreting, presenting), writing scientifically and evaluating sources. In an independent research project, students are expected to design, conduct and present research in which they examine the effect of an environmental factor of their choice on plant growth. Most draw on home-based practices of growing plants, and their projects usually include the use of grey water or some form of organic waste. Assessment criteria for the proposal include such aspects as using an appropriate experimental design (taking into account randomization, replication and control of local conditions) and developing a logical argument (with the hypothesis being supported by information from the literature). These criteria represent stronger epistemic relations (ER+).

These scientific literacies concepts and procedures have relevance in the disciplinary work as well. For example, the principles of experimental design could apply equally in chemistry and human kinetics and ergonomics experiments. However, the analysis indicates that they are addressed initially in ISCM tutorials in fairly generic ways, and only later drawn on specifically in the disciplinary work.

In terms of social relations, the required dispositional attributes and values associated with this experimental work are closely linked to the specialized procedures. Students are expected to work in rigorous, reliable, accurate, precise, honest, curious, logical, analytical, critical and evaluative ways, all attributes required of scientists (Matthews 2015). While these attributes might at first glance suggest stronger social relations, they in fact represent weaker social relations (SR–): their *focus* may be knower attributes but their *basis* for achievement are the epistemic relations outlined above. For example, as indicated in Table 8.1, honesty in collecting and presentation of data is emphasized in tutorial discussions, but since this is difficult to ascertain students are instead awarded marks based solely on the presented outcomes of their experiment. Furthermore, students are generally expected to work objectively in the sciences, which means they must suspend personal biases and subjectivity in favour of what the empirical data tells them. This valuing of objectivity in the sciences therefore also represents weaker social relations (SR–).

Student response to the knowledge code

In terms of the science-related knowledge code, when students exhibit poor realization in the disciplinary component their answers are usually conceptually incorrect, too generalized or oversimplified (see Ellery 2017b). In other words, they recognize that a knowledge code is necessitated but are unable to produce the required cognitive or abstract standard. Similar code recognition but poor code realization is evident in the scientific literacies work and is usually linked to, amongst other things, students not being acceptably accurate or precise in their measurements, logical in their argument or objective in gathering their data. Therefore, although social relations are relatively weaker in this code, it appears that students need to pay some attention to the *valuing* of procedures as they influence being able to produce the requisite text. The student interviews indicate that educational background may play a role in poor realization of the science-related knowledge code at a university level. As one student commented:

TABLE 8.1 Specific translation device for knowledge code in ISCM (adapted from Ellery 2016, 2017a)

Code	Indicators	Empirical evidence from observations, document analysis and staff interviews
Science-related knowledge code	**ER++** Disciplinary component Disciplinary conceptual and procedural knowledge is emphasized while	**Concepts** (from lecture handouts) Geology: processes of river formation Chemistry: concept of a mole Physics: process of radioactive decay HKE: factors that affect range of joint motion **Procedures** (from practical handouts) Plot river course on topographic map Titrate a solution to determine molar concentration Calculate a decay constant. Measure range of joint motion using a goniometer **Assessment** (test): State main river formation processes; explain process of subduction in plate tectonics
	SR– dispositions and opinions are downplayed	**Dispositional attributes and opinions** **Assessment** (tutorial discussion): Opinion sought on nuclear energy as a form of energy generation, but answers need scientific base **Assessment** (field trip) Stated focus is careful, rigorous and systematic observation, but marks allocated for correct identification of geological features
	ER+ Scientific literacies component Scientific literacies conceptual and procedural knowledge valued while dispositions and values are downplayed	**Concepts** (from tutorial handouts) How scientific knowledge is generated (measurement, observation, inductively, deductively, developing hypotheses, conducting experiments, making predictions) The basis upon which knowledge claims are made (use of empirical data and models of reality; recognizing tentative nature of science and idealization of laws) **Procedures** (from tutorial handouts) Design and conduct experiments Develop coherent arguments Work with data Write scientifically Evaluate sources **Assessment** criteria in a proposal: Experimental design (principles of randomization, replication, control of conditions); Logical argument (hypothesis supported by literature)
	SR–	**Dispositional attributes** Tutorial discussion: On the value of and need for honesty in reporting in experiment, but assessment criteria based on final outcome of experiment

I won't lie, it's the first subject that I can say it was quite difficult for me … last year when I was doing my matric, I didn't see how deep is science, but in ISCM I saw how deep is science … in terms of being critical and put your understanding towards your work.[3]

(Kanelo)

Academic practices-related knower code

Epistemic relations and social relations

The second code legitimated in ISCM is an academic practices-related knower code. The term 'academic practices', which has social connotations and infers underpinning values as opposed to the more acontextual 'study skills', is used here to describe this work. It relates to what staff perceive to be the main purpose of the course: enabling students to become effective learners in a higher education science context. This purpose arises from a concern that once students leave the access programme and enter the mainstream, support for learning is either greatly reduced or non-existent. As one staff member stated in an interview: 'When they leave us students need to be able to get on with the job … work on their own' (Lecturer 1). Since there are no underpinning specialized concepts, but instead students are required to engage in and develop learning practices appropriate for a higher education context, it embodies relatively weaker epistemic relations (ER–).

As indicated in Table 8.2, epistemic relations here are primarily procedural and linked to particular learning practices:

- organizational practices that relate to issues such as managing time and organizing notes;
- technical practices that are linked to their accessing information and taking good lecture notes from which they can learn;
- study practices that require students to prepare for and attend lectures, ask questions, review and consolidate work, practice calculations, and engage with and respond appropriately to feedback; and
- assessment techniques that include managing time in assessment and unpacking questions.

Working effectively in these academic practices is contingent on understanding the contextual 'rules', as outlined in Table 8.2. For example, in order to develop the practice of reviewing and consolidating lectures, students need to know that the ISCM context requires them to work constructively and independently outside class, otherwise it is unlikely they will be successful. Likewise, answering test questions effectively requires understanding that verbs of instruction (describe, explain, evaluate) determine the kind of answer, and that the supporting text and mark allocation signal the scope of answer required.

TABLE 8.2 Specific translation device for knower code in ISCM (adapted from Ellery 2016, 2018)

Code	Indicators	Empirical evidence from observations, document analysis and staff interviews
Academic practices-related knower code	Dispositions and behavioural attributes for shaping own learning are emphasized and valued	**ER−** **Contextual 'rules'** (from tutorial interactions) Organizing principles of library, Internet, textbooks, dictionary, words (suffix, prefix), learning context, assessment questions — **Procedures** (from tutorial handouts) Organizational practices: manage time, file notes; Technical practices: access information and take lecture notes; Study practices: prepare for and attend lectures, ask questions, review and consolidate work, practice calculations, work with feedback; Assessment techniques: manage time, unpack questions
		SR+ **Dispositional attributes** Staff interview quotes: 'realising they don't know; study effectively and independantly; developing deeper and better understanding; right kind of academic level; willingness to engage; active participation'

It is through these academic practices that a particular kind of learner is being legitimated. In interviews, as indicated in Table 8.2, staff articulate the norms and values they perceive as necessary to become an effective learner in ISCM in particular and in a higher education context in general. All five staff interviewed spoke about learner independence. As one stated: '[They] need to become capable as students to study effectively and independently as we cannot always be here to support them' (Lecturer 5). In this regard another said: '[They should] not always rely on someone to teach them everything nor rely on someone to check whether they have understood' (Lecturer 1). Most staff emphasized the need for students to develop proper understanding. As one mentioned: 'It's not just surface content … it's about developing a deeper and better understanding … [they] cannot just rote learn all the time' (Lecturer 3). They also spoke about engaging at a higher level than at school. As one suggested: '[Staff] need to bring them to the right kind of academic level in terms of the scientific reasoning approach' (Lecturer 3). Most spoke about developing metacognitive, reflective and reflexive understanding. In this regard one stated: 'It's metacognitive thinking, you know. Realizing that they don't know what's going on, thinking about *how* they learn, and *what* they are going to learn, and *why* they are learning it' (Lecturer 2). Other aspects that staff spoke about were seeking help when needed, a willingness to engage and be challenged and active participation. In short, they view the academic practices work as engendering independence in learning and development of depth understanding, which is conceptualized here as students becoming and being *autonomous* learners. This invokes learners underpinned by relatively stronger social relations (SR+).

Autonomy in learning is enabled in ISCM through tutorial interactions in which appropriate learning practices such as preparatory reading for a topic, consolidating lectures and practical sessions, and active reflection on feedback for improvement, are supported and modelled in order that students can later do such activities independently of lecturer input. Outputs from these tutorial activities are not assessed directly, but many disciplinary and scientific literacies assessment tasks in ISCM are instead designed to not only assess knowledge of concepts and procedures but also to test autonomy in student engagement. Therefore, what may appear to be a question exhibiting distinctly stronger epistemic relations, such as the identification of a limiting reagent in chemistry, in fact has stronger learning-context social relations embedded in the task, based on how the topic has been addressed in class. If students have not heeded the cue to watch the necessary YouTube video, practiced calculating molar mass and number of moles and developed a proper understanding in order to answer a question that is phrased differently to that in class, it is highly unlikely that they will be successful in the task. In other words, the basis for achievement of the academic practices knower code is not measured directly, but rather indirectly through assessment tasks in the science-related knowledge code.

Student response to the knower code

When asked about poor performance in knowledge-code assessment tasks many students recognize the need to engage with their studies in particular ways, as demanded by the knower code, but are less successful at realizing such practice (see Ellery 2017b). In this regard students speak about the need to learn more: 'I just did not learn enough for this test, I will be honest' (Liwa). Along with insufficient engagement, how students engage is also an issue. Some appreciate the need to learn for understanding: 'I think I just did it [learning] for marks not for knowledge' (Khuselwa). Poor engagement often arises from a misreading of context requirements: 'I didn't know [learn] this because when she [the lecturer] answered that question she didn't spend a lot of time with it, she just browsed through it' (Anele). Students also recognize the need to practice technical tasks more: 'I understand the concept and method of calculating but I made a confusion in the test … I should practise more often to familiarize myself with the conversions' (Mbuyiselo).

In summary, students signal poor realization of the academic practices-related knower code in interviews when they speak of, amongst other things, ignoring difficult work, poor engagement with resources (including feedback), inconsistent study habits, reliance on rote learning and surface understanding, and dependence on an authority figure to direct their learning, which they acknowledge originates from their school contexts. These habits suggest unspecialized learning-context dispositions that exhibit weaker social relations (SR–). When combined with the weaker epistemic relations (ER–) of the generic procedures or skills being taught, this suggests that students enact this component of the course as a relativist code (ER–, SR–), instead of a knower code (ER–, SR+). There is thus a code clash for the learning context. Despite explicit articulation and strong support for the knower code in ISCM, students are not easily making the necessary shift. In this regard, students mention the challenge of changing entrenched study practices that have ensured success previously:

> I did well at school, very well, I was often first or close in my class … my pattern was to study very hard just before tests – sometimes for hours … [but here at university] I always run out of time 'coz there are tests and assignments and homework and stuff and it all takes too much time … I guess I should be working more consistently like every day like you tell us [laughs].

> (Mandisa)

Epistemological access in ISCM

Figure 8.2 indicates that ISCM legitimates two distinct codes. In summary, the science-related knowledge code focuses mainly on students understanding of disciplinary and scientific concepts and procedures and forms the primary basis for success in ISCM. While the disciplinary and scientific literacies components can be plotted separately on the specialization code plane based on their different strengths of epistemic relations, they still form part of the same code in which epistemic relations are emphasized and social relations downplayed. The academic practices-related knower code focuses primarily on student learning and relates to students

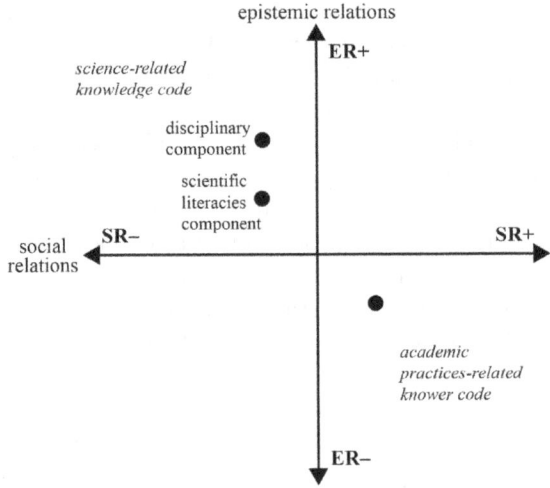

FIGURE 8.2 Specialization codes enacted in the ISCM course

becoming and being independent learners responsible for their own knowledge and understanding, which represents weaker epistemic relations and stronger social relations.

Because of the way the course and assessment practices are structured in ISCM, there is a close relationship between the two codes. In other words, if students do not work independently and develop deep understanding, as required by the academic practices-related knower code, they seldom can provide the depth that science answers require by the science-related knowledge code. This raises the key finding of this study that realizing the science-related knowledge code is contingent on students realizing the academic practices-related knower code. This hierarchy of access to these two codes has major implications for students' success and forms the basis for the more generalized conceptual model of epistemological access in the following section.

Epistemological access in higher education science

In serving its purpose of enabling access to the sciences ISCM has a dual focus of developing students' science knowledge as well as their dispositions as learners. ISCM is therefore unrepresentative of science higher education courses in general in that it has an explicit focus on developing students as learners through overt articulation, support, modelling and scaffolding of appropriate learning practices. However, many science-related higher education courses do, in fact, legitimate the kind of learner similar to that valued in ISCM (see Case 2013, Ellery 2017c). In other words, by working at a high volume and pace and addressing most concepts only once in class, they are legitimating a self-regulated learner who will work independently to ensure their own understanding. However, provision of support for such work is little or non-existent (Ellery 2017c). It appears that in current content-laden science-related curricula it is generally assumed that students will invoke learning approaches necessary for their success.

Since many science courses likely legitimate both a knowledge code (explicitly) and a knower code (implicitly), two levels of access are required for success in the sciences in general. The first level represents access to the specialized knowledge and procedures associated with stronger epistemic relations and is referred to as *science-context access*. This is the core of any science curriculum. However, since academic success is influenced by both the mastery of specialized knowledge *and* the 'capacity to be in control of one's own learning' (Edwards 2015: 14), access at the science-context level may well be constrained, as this study suggests, by poor access at the learning-context level. The focus here on becoming a particular kind of learner in a particular academic context is referred to as *learning-context access*. Without access at both levels, students are unlikely to be successful.

Access to two codes, with different strengths of epistemic relations and social relations, has implications for curriculum and pedagogy. The science context with its stronger epistemic relations poses certain structural constraints in any curriculum. To give an example, understanding the process of evolution as described in many higher education science curricula requires a prior understanding of base concepts such as the structure of DNA and the process of gene recombination. In contrast, the learning context with its stronger social relations demands a certain kind of learner as perceived relevant by staff. Aspects of the curriculum associated with the learning context are thus influenced by the purpose of the course and the values of the lecturers and are much more negotiable than those of the science context. In this regard, because ISCM is a science foundation course that is preparing students for success in the mainstream, an independent and self-regulated learner is legitimated. However, studies in engineering courses have shown that the ability to integrate multidisciplinary knowledge from a range of different contexts (Wolff and Luckett 2013) and the capacity to work with large volumes and at a high pace (Case 2013) may also (or instead) be key components of the learning context. Because there is more room to adjust expectations in learning-context aspects of the curriculum, I suggest this is where future effort needs to be focused to better accommodate students' backgrounds.

Prior educational experiences will influence access at both levels. Nonetheless, poor learning-context access in science-related disciplines is increasingly recognized as problematic (Adendorff and Lutz, 2009, Case 2013) and several studies have shown that prior educational experiences can limit uptake of, and access to, a new code (Hoadley 2007, Chen 2010, Pearce *et al.* 2015). It is suggested that the larger the articulation gap between secondary and higher education requirements, the more challenging the uptake (Scott 2012). The difficulty of changing dispositions and attitudes based on entrenched habits needs to be recognized, accommodated and supported in our educational practices (Pym and Kapp, 2011), without which certain groups of students will continue to be alienated and excluded from the system. From a social justice perspective, there is an obvious and urgent need to better support all students entering the higher sector today.

Conclusion

The contribution of this study, through enacting the concepts of specialization codes, has been the recognition of two levels of concern in a curriculum: the science (or disciplinary) context and the learning context. These two sets of concerns are always present in an educational field of practice, but empirical studies usually focus on either one or the other. By examining them together, these concepts allow a more holistic and nuanced analysis of curriculum and the different bases of achievement that are articulating within the educational context. As such, this approach offers the necessary subtleties to identify a key site of student difficulties with success – at the level of the learning-context knower. This highlights a somewhat unusual finding for education in the sciences, namely that learning-context social relations are key to enabling access to the highly valued science knowledge that will ensure success in academia. This emphasizes the need to significantly rethink science higher education to take into account methods for teaching both the knowledge of science and how students can be apprenticed into becoming science learners in a higher education context, thereby enabling epistemological access. This is particularly relevant for students whose prior socialization at home and school does not match well with expectations at university.

Notes

1 For the larger study to which this chapter relates, see Ellery (2016).
2 The literacies approach in ISCM is based on Street's (2006) ideological model that assumes there are multiple literacies and that these are acquired in socio-cultural contexts.
3 Pseudonyms are used for student quotes.

References

Archer, L., Hutchings, M. and Ross, A. (2003) *Higher Education and Social Class: Issues of Exclusion and Inclusion*, London: Routledge.

Adendorff, H. and Lutz, H. (2009) 'Factors influencing the learning process in first-year Chemistry', in B. Leibowitz, A. Van der Merwe, and S. Van Schalkwyk (eds) *Focus on First-Year Success*, Stellenbosch: Sun Media, 167–80.

Arum, R., Gamoran, A. and Shavit, Y. (2012) 'Expanded opportunities for all in global higher education systems', in L. Weis and N. Dolby (eds) *Social Class and Education: Global Perspectives,* New York: Routledge, 15–36.

Barnett, R. (2007) *A Will to Learn: Being a Student in an Age of Uncertainty,* Maidenhead, Berkshire: SRHE and Open University Press.

Bernstein, B. (2000) *Pedagogy, Symbolic Control and Identity: Theory, research, critique,* revised edition, Oxford: Rowman and Littlefield.

Case, J. M. (2013) *Researching Student Learning in Higher Education: A social realist approach,* London: Routledge.

Chen, R. T. (2010) '*Knowledge and Knowers in Online Learning: Investigating the Effects of Online Flexible Learning on Student Sojourners*', unpublished PhD thesis, University of Wollongong.

Council on Higher Education Report. (2013) *A Proposal for Undergraduate Curriculum Reform in South Africa: The Case for a Flexible Curriculum Structure,* Pretoria: CHE.

Dall'Alba, G. and Barnacle, R. (2007) 'An ontological turn for higher education', *Studies in Higher Education*, 32(6): 679–91.

Edwards, A. (2015) 'Designing tasks which engage learners with knowledge' in I. Thompson (Ed.) *Designing Tasks in Secondary Education: Enhancing subject Understanding and Student Engagement*, Abingdon, Oxon: Routledge, 13–27.

Ellery, K. (2016) '*Epistemological Access in a Science Foundation Course: A Social Realist Perspective*', unpublished PhD thesis, Rhodes University.

Ellery, K. (2017a) 'Conceptualising knowledge for access in the sciences: Academic development from a social realist perspective', *Higher Education*, 74: 915–31.

Ellery, K. (2017b) 'A code theory perspective on science access: Clashes and conflicts', *South African Journal of Higher Education*, 31(3): 82–98.

Ellery, K (2017c) 'Framing of transitional pedagogic practices in the sciences: Enabling access', *Teaching in Higher Education*, 22(8): 908–24.

Ellery, K. (2018) 'Legitimation of knowers in science: A close-up view'. *Journal of Education*, 71: 24–38.

Hoadley, U. K. (2007) 'The reproduction of social class inequalities through mathematics pedagogies in South African primary schools', *Journal of Curriculum Studies*, 39(6): 679–706.

Lamont, A. and Maton, K. (2008) 'Choosing music: Exploratory studies into the low uptake of music GCSE', *British Journal of Music Education* 25(3): 267–82.

Longden, B. (2006) 'An institutional response to changing student expectations and their impact on retention rates' *Journal of Higher Education Policy and Management*, 28(2): 173–87.

Matthews, M. (2015) *Science Teaching: The Contribution of History and Philosophy of Science* (2nd edn.), New York: Routledge.

Maton, K. (2014) *Knowledge and Knowers: Towards a Realist Sociology of Education*, London: Routledge.

Maton, K. (2016) 'Legitimation Code Theory: Building knowledge about knowledge-building' in K. Maton, S. Hood and S. Shay (eds) *Knowledge-building: Educational Studies in Legitimation Code Theory*, London: Routledge, 1–23.

Maton, K. and Chen, R. (2016) 'LCT and qualitative research: Creating a translation device for studying constructivist pedagogy', in K. Maton, S. Hood and S. Shay (eds) *Knowledge-Building: Educational Studies in Legitimation Code Theory*, London: Routledge, 27–48.

Maton, K. and Howard, S. K. (2016) 'LCT in mixed-methods research: Evolving an instrument for quantitative data', in K. Maton, S. Hood & S. Shay (eds) *Knowledge-Building: Educational Studies in Legitimation Code Theory*, London: Routledge, 49–71.

Morrow, W. (2007) *Learning to Teach in South Africa*, Pretoria: HSRC Press.

Morrow, W. (2009) *Bounds of Democracy: Epistemological Access in Higher Education*, Cape Town: Human Science Research Council Press. Retrieved from http://www.hsrcpress.ac.za

Muller, J. (2014) 'Every picture tells a story: Epistemological access and knowledge', *Education as Change*, 18(2): 255–69.

Muller, J. (2015) 'The future of knowledge and skills in science and technology higher education', *Higher Education*, 70: 409–16.

Pearce, H., Campbell, A., Craig, T. S., le Roux, P., Nathoo, K. and Vicatos, E. (2015) 'The articulation between mainstream and extended degree programmes in engineering at the University of Cape Town: Reflections and possibilities', *South African Journal of Higher Education*, 29(1): 150–63.

Pym, J. and Kapp, R. (2011) 'Harnessing agency: Towards a learning model for undergraduate students', *Studies in Higher Education*, 38(2): 272–84.

Science Extended Studies Programme Review Report (2011) *Curriculum Review: Science Extended Studies Programme, Rhodes University*, unpublished report, Rhodes University.

Scott, I. (2012) 'Alternative Access to University: Past, Present and Future', in R. Dhunpath and R. Vithal (eds) *Alternative Access to Higher Education: Underprepared Students or Underprepared Institutions?,* Cape Town: Pearson, 26–50.

Street, B. (2006) *Autonomous and ideological models of literacy: approaches from New Literacy Studies.* Retrieved from http:/www.philbu.net/media-anthropology/street_newliteracy.pdf.

Wolff, K. and Luckett, K. (2013) 'Integrating multidisciplinary engineering knowledge', *Teaching in Higher Education*, 18(1): 78–92.

9

THE RELATIONSHIP BETWEEN SPECIALIZED DISCIPLINARY KNOWLEDGE AND ITS APPLICATION IN THE WORLD

A case study in engineering design

Nicky Wolmarans

Introduction

Traditionally, professional education has been about the acquisition of a body of knowledge that graduates are expected to 'apply' in their professional practice after graduation. In this chapter I argue that a model of professions as the application of disciplinary knowledge is inadequate because it fails to take into account the complexity of the 'real world' to which the knowledge is applied. This in turn has led to curriculum design choices that fail to prepare graduates to work dialectically between the complexity of specialized disciplinary knowledge and the complexity of the world in which it is employed.

Engineering provides a rich case study. It is a profession that has positioned itself as a science-based discipline, founded on a canon of well-defined disciplinary subjects. Engineering curricula have tended to focus on the transmission and acquisition of scientific concepts and the relations between concepts, culminating in a final 'capstone' design project. This capstone design project is intended to integrate the specialized disciplinary knowledge acquired throughout the curriculum for application in a single 'real world' project. Most projects are set up to mimic the sorts of projects that engineering graduates are likely to encounter in professional practice (Froyd *et al.* 2012, Harris *et al.* 1994). They are intended to bridge the gap between engineering science and engineering practice. However, many students lack the skills to design when confronted with these design projects for the first time, even when they have successfully completed their engineering science courses (see Kotta 2011 for a detailed study of students' experiences of senior design projects).

Over the last century of engineering education reform (Froyd *et al.* 2012, Grinter 1955, Mann 1918), employers have been calling for improved interpersonal and enabling skills, a strong foundation in the fundamental sciences, and the centrality of design in the curriculum. Recent studies of employer perceptions of graduate engineers report an improvement in, for example, teamwork, communication skills and management compared to the past (J. King 2007, R. King 2008). However, many engineering graduates still appear to be unable to apply scientific knowledge to solve professional problems; as J. King (2007: 7) reports:

> Although industry is generally satisfied with the current quality of graduate engineers it regards the ability to apply theoretical knowledge to real industrial problems as the single most desirable attribute in new recruits. But this ability has become rarer in recent years

In this chapter I address the question of what it means to 'apply theoretical knowledge' to 'real industrial problems' emergent from the complexity of the world. Using engineering design projects in the curriculum as a proxy for real professional problems, I present an analysis of the relationship between material artifacts and the abstract concepts used to analyze and mathematically model them for the purpose of design. The chapter draws on part of a larger PhD study (Wolmarans 2017a) in which 17 engineering design projects located three engineering streams where investigated. Only two of the projects are presented here, both introductory design projects. One is the first project in a sequence of civil engineering projects, the other is the first project in a sequence of structural engineering projects. A comparison of the two projects shows that different ways of simplifying 'real' projects for the purpose of learning have vastly different effects on the nature of the required reasoning.

I bring together concepts from the Semantics and Specialization dimensions of Legitimation Code Theory (Maton 2013, 2014, 2020), because of their capacity when integrated to analyze relationships between objects of knowledge and knowledge of objects. The concept of *semantic gravity* provides a lens to analyze the significance of knowledge of the object of design, while the concept of *semantic density* provides a lens to analyze the complexity of the reasoning. By coding the knowledge requirements of each step in the design thinking process, I am able to show shifts between knowledge of complex things (using the concept of *ontic relations*) and knowledge of complex theoretical concepts (using the concept of *discursive relations*). The analysis contributes to building a more robust model of professional reasoning which has implications for learning. The study provides insight into the limitations of certain types of tasks when they constrain the complexity of thinking about the 'things' being analyzed. In short, privileging specialized knowledge over knowledge of objects effectively distorts the complex dialectical relations involved in professional reasoning. Although the case presented in this chapter is that of engineering, these findings have implications for a range of different professions.

Theoretical framework

The theoretical backdrop to this study begins from Bernstein's distinction (2000) between 'singulars' and 'regions'. He described "singulars as bodies of disciplinary educational knowledge that are strongly bounded from other educational knowledge and from external concerns in the world beyond education. When discussed in broad terms, disciplines such as Physics, Mathematics and History might be described as singulars, each with its own specialized concepts and rules of conceptual relations among concepts. Bernstein described 'regions' as the interface between singulars and the practical concerns of the world – they involve selection of ideas from a range of singulars and their application to a field of practice beyond education. Professions are usually described as regions. This model of knowledge assumes singulars precede regions which in turn project that knowledge onto the world (see Smit 2017 for a detailed critique of this notion of singulars and regions in a study of thermodynamics in science and engineering).

Bernstein's model is useful in that it does recognize a distinction between the structure and organization of specialized disciplinary knowledge (e.g., fluid mechanics) and the structure and organization of knowledge in engineering practice where that knowledge is used in relation to the design of complex systems. However, the model has significant shortcomings. For one thing, it does not address the nature of the relationship between knowledge inside the discipline (abstract fluid mechanics) and knowledge outside the discipline (the design of a slipway for a new yacht basin) – see, for example, Wolmarans (2017b). Based on Bernstein's model of regions, scholars have characterized professional education in terms of mastery of specialized disciplinary knowledge *prior* to its application to problems in the world – the 'word' before the 'world' (Beck 2002, Beck and Young 2005). This model tacitly assumes that if graduates have learned the concepts and legitimate rules of combination among concepts within a disciplinary specialization, then such theoretical knowledge can be unproblematically 'applied' to external problems. It is a model of knowledge that tacitly underpins many engineering programs. But one of the under-developed aspects of this notion of 'regions' is that it leaves us blind to the significance of the contextual detail inherent in professional problems emergent from the world. It fails to take account of the need to translate knowledge structured by the logic of internal conceptual coherence into a logic structured by the external realities of the world. In short, it fails to explore the nature of the external problem itself.

Legitimation Code Theory (LCT) includes a number of conceptual tools that offer a more nuanced way of viewing knowledge practices (Maton 2014). The Semantics dimension of LCT is a particularly attractive analytic tool for investigating professional knowledge in terms of the relations between 'abstract' specialized disciplinary knowledge and 'contextually embedded' professional problems. Semantics centres on exploring the organizing principles underlying practices in terms of *semantic gravity* (SG), which conceptualizes degrees of context-dependence, and *semantic density* (SD), which conceptualizes complexity. Both semantic gravity and semantic density describe a range of relative strengths from stronger to weaker. For example, when SG is weaker, the underlying practices are relatively less dependent on any specific context, while stronger SG means that the underlying practices are relatively more dependent on a specific context. In the case of SD, more complex practices exhibit relatively stronger SD and less complex practices exhibit relatively weaker SD.

For the purposes of this study:

- stronger semantic gravity (SG+) means that making sense of the problem itself and the specialized knowledge recruited to develop a solution are strongly dependent on the specifics of the problem;
- weaker semantic gravity (SG−) suggests that the specialized knowledge is generalizable across multiple contexts;
- stronger semantic density (SD+) indicates projects requiring more complex reasoning; and
- weaker semantic density (SD−) indicates a simpler problem.

SG and SD vary independently of each other and can shift strength through the duration of any activity. SG↑ indicates a process of strengthening semantic gravity, in this case increasing the specificity of a project. SD↓ indicates weakening semantic density, here simplifying either relations among specialized concepts or the aspects of the object of design to be considered.

Semantic codes have been used productively to show the importance of shifting between more abstracted theory (SG−) and more concrete examples (SG+), both in driving cumulative knowledge-building and to identify tacit evaluative criteria. See, for example, Blackie (2014) on Chemistry, Georgiou et al. (2014) on Physics, Maton (2013) on Biology and History, and Shay and Steyn (2016) on Design. In terms of recruiting specialized knowledge to solve professional problems, semantic gravity can provide a means of describing the relation between the 'abstract' theoretical knowledge in its academic form and the more 'concrete' problems that emerge from practice. When looking at professional education, semantic density offers a way to look at progression through a curriculum based on increasing levels of complexity, or strengthening semantic density. Shay and Steyn (2016) used semantic codes to analyze a sequence of projects in an introductory design course. Building from their results, they were able to redesign the sequence of tasks into a coherent trajectory of increasing complexity and to identify the link between complexity and 'concreteness' of the project.

I have also used Semantics as an analytical lens. However, semantic density has not yet been used in LCT studies to distinguish between i) the complexity of the object of study or the artifact of design (the thing, what it is and how it works and ii) the complexity of the knowledge recruited to do the analysis (the specialized disciplinary knowledge). In order to make this distinction I draw on the LCT dimension of Specialization, specifically the distinction that Maton (2014) makes between *ontic relations* between knowledge and its objects of study and *discursive relations* between knowledge and other knowledges. For my study, this can be enacted to distinguish between those concepts with which we make sense of things in and of the world (*ontic relations*) and those concepts defined within any specific discipline that conform to disciplinary rules (*discursive relations*). For example, the way in which we understand a bicycle, what it is used for, how to ride it, what it looks like, requires knowledge of a bicycle in terms of its physicality and one's experience of seeing or using a bicycle (ontic relations). Modelling the dynamics of motion, the principles of friction between road and tire, and the strength of the frame requires specialized disciplinary knowledge (discursive relations). In this study I have analytically separated:

- the complexity of the disciplinary knowledge recruited in the task – semantic density of discursive relations or DSD; and
- the complexity of the knowledge of the object of design – semantic density of ontic relations or OSD.

Put simply, I have separately analyzed the complexity of the object (OSD) and the complexity of the specialized knowledge recruited (DSD).

The case study

The data presented in this chapter draws on a PhD study of 'The nature of professional reasoning' (Wolmarans, 2017a). The study was based on an analysis of three sequences of engineering design projects located in two engineering degree programs. Engineering was chosen as the case study because it is a profession that is founded on a well-established scientific knowledge base. The knowledge tends to be more explicit and clearly bounded than in some of the newer professions or those based on social sciences and humanities. The reason for investigating engineering design projects located in a curriculum rather than in practice is twofold. First, the assessment requirements in an educational task require that the design reasoning is elaborated and made more explicit than might be the case in professional practice where professionals may draw more tacitly on their specialized knowledge. Second, because the data was collected in an educational context, engineering design projects were selected because engineering design is seen as the bridging subject between knowledge and practice. Engineering design projects are usually intended to mimic professional engineering projects, making them a reasonable proxy for investigating professional knowledge in action.

Some of the projects analyzed in the broader study include, for example – the design of a culvert to attenuate floodwaters on a particular watercourse; the design of a multistory parking garage; and the specification of the requirements for a power station. Although all the projects in the study were identified as 'design' projects by the lecturers concerned, they did take different forms with different educational objectives and privileged different forms of knowledge. This provided an opportunity to investigate the effect that different ways of simplifying a professional project (recontextualizing choices) have on the nature of the required reasoning.

In this chapter I present a comparison of two of the projects: the design of a bikeshare scheme (U1: Bikeshare scheme) and the analysis of the loading on a structure (S1: Parking structure). Both are introductory projects, the first in a learning sequence of five design projects. Both are intended to mimic aspects of the sorts of projects that professional engineers encounter in their professional practice. However, because the projects are located early in the curriculum, they require substantial simplification in comparison to 'real' engineering projects. The projects need to be constructed so that they are appropriate for students not quite midway through their curriculum, with limited exposure to the range and complexity of the engineering sciences. The projects are further constrained by the limited time available in the curriculum. These two projects were selected because they represent recontextualization based on very different principles of recontextualization. In the case of U1 (Bikeshare scheme) the disciplinary knowledge needed to understand the project requirements was reduced (although it was reintroduced later in the project). In the case of S1 (parking structure) the structure itself was simplified. The analysis is relevant to understanding the consequences of teaching engineering sciences in relation to simplified idealized objects and illustrates some of the unintended consequences of the choices made in terms of how design tasks are simplified.

Data

The data collected for each design project included the design brief and a design solution. The design brief is provided to students at the beginning of the project and lays out the project requirements and task instructions. Briefs tend to identify a problem scenario or client need and describe or refer to a context in which the need or problem arises and in which the proposed design solution must operate. The design solution describes the proposed artifact and documents evidence of the proposed artifact's performance in context. The design solution collected was in the form of either a solution memorandum (a sample solution prepared by the lecturer, a 'model answer') or a student design report. Because design rarely has a 'model answer' and marking rubrics are inadequate to show the details of design reasoning, in most cases student solutions were used as a proxy for a solution memorandum. In these cases, a 'good' student design solution was selected for analysis. The basis of selection was on the grade achieved by the student for the particular project.

Four units of analysis were identified for each project. The design brief typically prescribed an artifact to be designed (*artifact prescribed*) and described or identified a context in which the artifact needed to perform (*context described*). The design solution was analyzed in terms of the resultant artifact proposal (*solution specified*) as well as the process of reasoning required to move from the brief to the final solution (*inferential reasoning*). These four units of analysis – the *artifact prescribed*, the *context described*, the *solution specified,* and the *inferential reasoning* – were coded for each design project. For clarity, the four units of analysis described above are summarized

TABLE 9.1 Units of analysis

Data	Unit	Description of each unit of analysis
Design Brief	*Artifact prescribed*	Each design brief usually prescribed an artifact to be designed. In some cases, the required artifact may be left to the discretion of the students, emergent from the design purpose.
	Context described	All artifacts function in a context, and the context places requirements and limitations on the design of the artifact.
Design Solution	*Inferential reasoning*	The inferential reasoning refers to the ideas generated, analytical models of potential performance and decisions made during the process of design, from brief to solution.
	Solution specified	At the end of the process of design a final solution artifact is specified. Typically, a design solution would be in the form of a set of technical drawings detailing the artifact proposed as a solution to the design problem.

in Table 9.1. The results of the analysis of each of the units of analysis for the two projects are shown in Table 9.4 (U1: Bikeshare scheme) and Table 9.5 (S1: Parking structure), further below.

The two projects presented, U1 (Bikeshare scheme) and S1 (Parking structure) are both considered to be the first design project in a sequential trajectory of design projects. These projects are compared because they illustrate two significantly different recontextualizing principles evident in the briefs. The parking structure (S1) is recontextualized by simplifying the artifact significantly (OSD↓). The bikeshare scheme (U1) is recontextualized by reducing the disciplinary knowledge required (DSD↓).

Analysis and discussion

LCT Semantics was used to investigate the nature of the reasoning in engineering design projects. Semantic gravity (SG) was used to analyze relations between ideas and the object that they describe, and semantic density (SD) was used to investigate the relation between concepts. In terms of semantic density, it became necessary to distinguish relations between formal theoretical concepts defined within specialized disciplines from more 'everyday' concepts used to make sense of the object of design and the context in which it was intended to operate. This resulted in the distinction between semantic density of discursive relations (DSD) and semantic density of ontic relations (OSD).[1]

The analysis is presented below in three stages. The first stage describes the development and explanation of how semantic gravity and semantic density were

TABLE 9.2 Semantic gravity categories of analysis

SG+/–	Code description	Examples
Generic or idealized, described in terms imposed from a body of disciplinary knowledge.		
SG– –	Meaning resides in **general** laws or concepts that **transcend** contexts; it does not require a concrete reference to make sense.	S1 solution: A compressive force which is transferable across all contexts and is applicable to all bodies as a result of gravity.
SG–	Meaning is **imposed** from a disciplinary body of conceptually coherent knowledge, but the general law/s or concept/s are **specialized** for application to an object or class of object.	S1 artifact: The structure is idealized but also unrealistic and impractical. It is designed to elicit analytic techniques rather than to perform a material functional.
Specific and detailed, described in terms emergent from the context		
SG+	Meaning relates to (originates in) a **type** of object or system; it refers to real objects, but abstracted from a specific, unique instance to a **class/type** of object.	U1 artifact: A type of system; the brief does not prescribe the specifics of any particular system.
SG++	Meaning relates directly to a **specific instantiation** of an object, described in **rich detail** specific to a **unique** case/situation.	U1 context: A specific campus accessible to students as part of the design project. Familiarity with the specifics of the campus is required.

operationalized in this study. The tables summarizing the categories that were used to analyze the data (Table 9.2 and 9.3) make reference to examples from the data. The examples are presented in the second stage where analysis of the two projects is compared. The process of reasoning required to develop a solution to the task is further elaborated in the third stage, where each step in reasoning is linked in a chain or network of inferential steps.

Operationalizing semantic gravity and semantic density

The analytical categories used for semantic gravity are shown in Table 9.2. At the first level stronger semantic gravity is distinguished from weaker semantic gravity based on whether meaning emerges from an understanding of the contextual or material detail (SG++ and SG+) or appears to be imposed from a specialized body of knowledge (SG– – and SG–). Within stronger semantic gravity I differentiate between *specific* or unique artifacts or contexts (SG++) and somewhat more general *types* or classes of artifacts or contexts (SG+). Within weaker semantic gravity I distinguish between *generalized* laws or principles that transcend contexts (SG– –) and generalized theories *specialized* and imposed on the artifact or context in order to describe it (SG–).

TABLE 9.3 Semantic density categories of analysis

SD+/−	Description	Examples
	Integrated or condensed into a coherent whole	
SD++	Multiple **interdependent** concepts/components integrated into a coherent whole. Theoretical antecedents/ causal interdependencies are **embedded** and not identified explicitly	DSD++: As entry projects neither required students to identify and select relevant disciplines from others. OSD++ U1 context: Students are referred to the UCT campus; a complex context with emergent characteristics.
SD+	Multiple **interdependent** concepts/components integrated into a coherent whole. However, relevant disciplines/concepts/objects are explicitly **identified** while retaining **simultaneous** interdependencies	DSD+ S1 inference: The calculations integrate geometric considerations, mathematical techniques and structural analysis. OSD+ U1 inferences: Simultaneous consideration of surveyed route, gradients, chosen bicycles, inexpert users, and shared user spaces.
	Separated or elaborated into constituent parts	
SD−	Relevant concepts/components are identified and **separated** into a **linear sequence** of prescribed relations.	DSD− S1 artifact: The structure is designed to demonstrate a sequence of analytical techniques at the expense of functionality. OSD− U1 solution: The solution is presented as a sequence of features listed in the report.
SD− −	A **single** concept/component is identified as relevant and is **isolated** and **dislocated** from its disciplinary/contextual relations.	DSD− − U1 context: Located in a surveying course identifies surveying knowledge as the only relevant discipline. OSD− − S1 inference: Each structural element is analyzed in isolation from the others.
SD_0	No specialized concepts or knowledge of the object are required.	DSD_0 U1 artifact: Students are referred to Wikipedia for a description of a bikeshare scheme. OSD_0 S1 context: The structure is stripped of interaction with a real context.

The analytical categories used for semantic density are shown in Table 9.3. Ontic semantic density (OSD) here refers to concepts used to make sense of the 'things' of the design, where coherence among concepts tends to be held together by the contextual details of the material artifacts – what they are and how they work. Discursive semantic density (DSD) here refers to formal concepts defined within a specialized body of knowledge and the formal conceptually coherent relations among them. As described previously, OSD requires knowledge of the physicality and experience of an objects, while DSD requires knowledge of the formal specialized knowledge used to analyze the object.

The strength of semantic density was categorized in relation to the level of detail, number of concepts and number of relations between the concepts. It relates to the nature of those conceptual relations needed to complete the design task. Although OSD was differentiated from DSD, the principles that define the coding categories are the same. In both cases stronger semantic density (SD++ or SD+) implies integration as the main principle of categorization. Weaker semantic density (SD− − or SD−) implies separation or elaboration as the main principle of categorization.

Within stronger semantic density, SD++ means that integrated interdependencies are embedded but not necessarily obvious. For example, students may need to identify appropriate disciplinary knowledge or concepts within a discipline without being given direction, or students may need to distinguish between those aspects of an artifact which are relevant to the design from those that are incidental to the task. SD+ indicates that the brief or instructions are given in such a way that the relevant components or concepts have been identified for students, although the parts retain necessary simultaneous interdependence. For example, the brief may instruct students to design a gearbox, including selecting a motor, bearings and power transmission elements and designing the shaft. This identifies separate components for students, but each component interacts interdependently with the others. All design decisions interact with other design decisions.

Within weaker semantic density, SD− means that each component or concept has been identified for students and retain a sequential relation. In other words, each relevant concept is identified and one concept links sequentially to the next concept. SD− − indicates that a relevant component or concept has been identified for students but separated to the point of dislocation from other components or concepts. For example, students may need to size a bearing dislocated from other machine components or students may be required to do a single calculation dislocated from the complex conceptual network of meaning defined within a discipline.

Comparative analysis of the two introductory design projects

The two projects compared in this chapter both represent design projects at the beginning of a sequence of design projects. At this point in the curriculum students have little to no experience in design and have limited proficiency in the engineering sciences but are expected to be relatively proficient in mathematics and the basic sciences. Thus, design projects need to be significantly recontextualized in comparison to 'real' professional projects, to align with the expertise that students are expected to bring. The first design project in the 'civils' stream in civil engineering required students to design a bikeshare scheme for the University of Cape Town Campus (U1: Bikeshare scheme). The project was a two-week block course in surveying. The project illustrates recontextualization that leaves the contextual and material details (ontic relations) intact, while reducing the disciplinary knowledge (discursive relations) to a single discipline with a sequence of specialized procedures prescribed. The first design project in the 'structures' stream in civil engineering required students to complete a loading analysis of a generic parking structure (S1: Parking structure). The project was integrated into the first structural engineering course and ran parallel to the engineering science content taught in the course. The project illustrates recontextualization based on stripping the contextual and material details (ontic relations) in order to develop specialized conceptual relations (discursive relations).

The analysis of the projects first compares the four units of analysis for each project, namely the *artifact prescribed*, the *context described* in the brief and the *solution specified* and the *inferential reasoning* required to develop the solution. The analysis illustrates the effect of recontextualizing choices on the complexity of understanding required to engage in the task. The categorization of the *inferential reasoning* is elaborated in the second analytical section. It shows the inferential chains of reasoning required of students as they develop an adequate design solution proposal. The analysis of the detailed *inferential reasoning* is particularly interesting in that it shows how significant stronger semantic density of ontic relations (OSD) are to the development of professional reasoning.

U1: Bikeshare scheme

The *artifact prescribed* in the brief was a 'bikeshare' scheme, a type of system (SG+). Initially the artifact can be understood in 'common sense' ways without recourse to specialized disciplinary concepts (DSD_0); students are referred to Wikipedia for a generic description of a bikeshare scheme. The information provided by Wikipedia functions to assist students to identify the relevant elements of the system and recognize how they interact as a system (OSD+).

The *context described* in this project is integral to the design. Rather than describing the context, students are referred to the UCT campus in its current form, including the terrain and usage: a congested campus with limited parking and narrow pathways shared by motor vehicles, busses and pedestrians. The climate is dry hot summers with strong winds and cool wet winters. The campus is built on the slope of a mountain and has many staircases and steep inclines. These details (SG++) are not specified in the brief and it is expected that the students will draw on their own experiences of the campus in the design. The design brief specifically identifies the campus and does make reference to some of the aspects of the campus context that might be considered, such as drawing attention to the importance of surrounding buildings and their usage to determine potential demand; potential interaction with the student bus service; and implied relations to existing roads for access. However, much of the detail is left to students to elaborate from their own everyday and embedded experience of the campus. Students need to either identify as significant or discard as irrelevant to the design details from their experience. The context is thus richly detailed and complex but without much guidance in terms of what to consider and what to ignore (OSD++). Initially students are able to engage with and make sense of the context without specialized disciplinary knowledge, although the location of the course in a surveying course does imply the importance of a single discipline over others that may be relevant to this project (DSD– –).

The *solution specified* has been analyzed in two parts because of the differences in the nature of the parts. The description bikeshare system (U1A: Bikeshare description) included bicycle selection, safety equipment, exchange logistics, storage and maintenance plans specific to this particular context (SG++). It was descriptive in nature and retained its unspecialized format (DSD– –). Each element was identified and described sequentially (OSD–) without much evidence of simultaneous interferences evident in the specification of the solution.

However, the real focus of the project was on the survey of the route and the modifications required to accommodate its purpose (U1B: Bikeshare route). This part of the solution was presented in the form of a digital elevation model (DEM), a specialized representation of a particular terrain (SG++). A number of vertical profiles at significant points along the route needed to be modified. Modifications were required along the route to accommodate, for example, inexpert users and interaction between multiple modes of transport. Although the process of developing the design (*inferential reasoning*) required networks of simultaneous considerations, the presentation of the solution is simplified. Each part of the map can be read as a sequence of positions along a contour map, and a sequence of vertical profiles (DSD–). And the route is a significant simplification of the campus, with each modification understood in relation to the terrain in one sense and the users in another (OSD–).

The *inferential reasoning* on the other hand required simultaneous consideration of the contextual details of the route in relation to specialized surveying knowledge. The network of inferences involved a range of different strengths of both OSD and DSD. Although a sequence of steps for the survey was provided in the brief, each step involved interrelations between measurements with specialized equipment, conversion into specialized surveying representations using multiple mathematical relations, and the separation of vertical and horizontal interactions and conversion into a map (DSD+). Once identified, the route itself could be considered sequentially, but at each point in terms of interactions between terrain and a range of users, including expected expertise of cyclists, interaction between different modes of users (cyclists pedestrians and motorized transport) (OSD+). The network of simultaneous reasoning shown in Figure 9.3 illustrates the nature of OSD+, DSD+. Table 9.4 is a summary of the analysis above and is illustrated in Figure 9.2.

S1: Parking structure

The *artifact prescribed* is a generic structure, stripped of all functional elements (ramps, lifts, parking and vehicle flow arrangements). Figure 9.1 shows the structure reduced to a collection of columns, beams and slabs configured in such a way as to offer a range of different analytical challenges. That is, disciplinary analytical requirements are imposed on the structure at the expense of functionality (SG–). As a result, each structural element can be treated independently of the others (OSD– –). As with

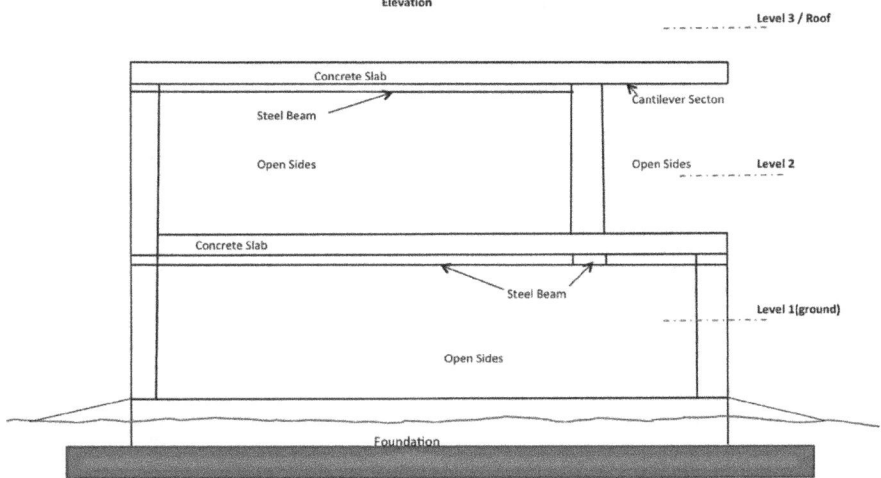

FIGURE 9.1 Idealized parking structure (S1)

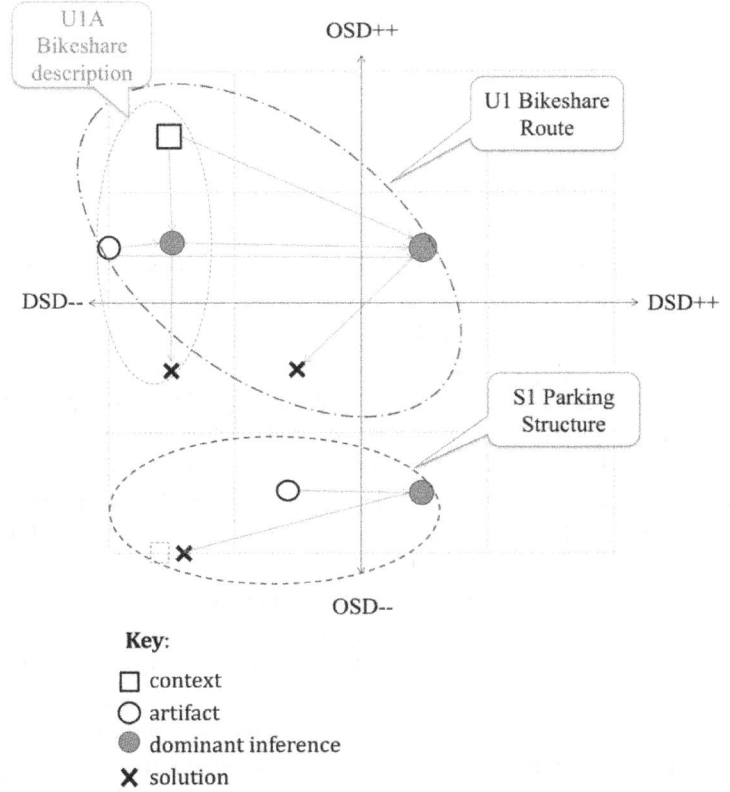

FIGURE 9.2 Effect of recontextualization by weakening OSD (S1) or DSD (U1)

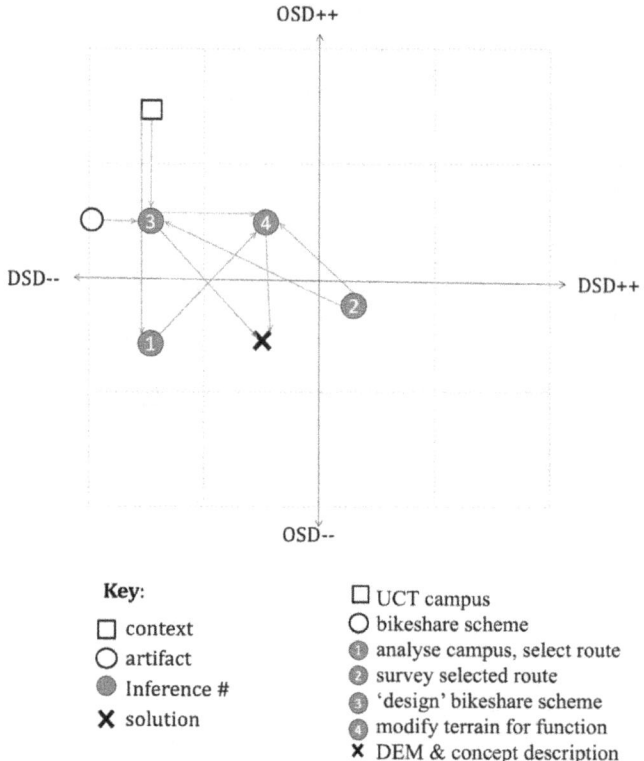

Key:

- □ context
- ○ artifact
- ● Inference #
- ✗ solution

- □ UCT campus
- ○ bikeshare scheme
- ① analyse campus, select route
- ② survey selected route
- ③ 'design' bikeshare scheme
- ④ modify terrain for function
- ✗ DEM & concept description

FIGURE 9.3 Inferential reasoning: U1 (Bikeshare scheme)

U1 (Bikeshare scheme) the project is located in a disciplinary course in structural analysis, signaling structural analysis as the only significant discipline. However, the description of the structure does include a list of symbolic markers used to link the elements through a sequence of disciplinary calculations (DSD−).

The *context* in which the structure operates is stripped from the project to the point of being completely general (SG− −) and replaced by a single symbolic representation of a 'live load' (variable load imposed under operation, specified for typical loading types). Even this level of understanding of a live load is not required to complete the task (OSD$_o$). The parking load is quantified as a uniformly distributed load given in the brief as 4kN/m^2 (DSD− −).

The *solution specified* was in the form of four compressive forces operating at the base of each of the four bottom columns. The form of the solution is completely transferable across all contexts, and in an abstract sense is applicable to all bodies in that all bodies on earth interact with a surface by a compressive force at their base as a result of gravity (SG− −). As with the *context description* given in the brief, the answer is a single symbolic term dislocated from its relation to other terms disciplinary terms (DSD− −) and stripped of any interaction in the world (OSD$_o$). Students are able to present the answer without engaging at all with what it means in the world.

TABLE 9.4 Summary of analysis U1: Bikeshare scheme

Unit of analysis	Semantic code
Artifact prescribed	Bikeshare Scheme: SG+, OSD+, DSD_0
Context described	University campus: SG++, OSD++, DSD– –
Solution specified	Bikeshare description: SG++, OSD–, DSD– –
	Bikeshare Route (DEM): SG++, OSD–, DSD–
Inferential reasoning	Bikeshare description: SG++, OSD+, DSD– –
(dominant mode)	Bikeshare Route (DEM): SG++, OSD+, DSD+

TABLE 9.5 Summary of analysis S1: Parking structure

Unit of analysis	Semantic code
Artifact prescribed	Generic structure: SG–, OSD– –, DSD–
Context described	Uniform distributed load: SG– –, OSD_0, DSD– –
Solution specified	Discrete compressive load: SG--, OSD_0, DSD– –
Inferential reasoning	Prescribed sequence of procedural calculations:
(dominant mode)	SG–, OSD– –, DSD+

The details of the analysis of the *inferential reasoning* are shown in Figure 9.4. The reasoning remains discursive, imposed on the structure for the purpose of structural analysis (SG–). Each element is considered sequentially but can be considered independently of the other elements (OSD– –), linked only through a sequence of calculations prescribed in the brief. Although the inferential steps are a prescribed sequence of calculations, each calculation does integrate a number of structural engineering principles that work together interdependently (DSD+).

The analysis summarized in Tables 9.4 and 9.5 is illustrated in Figure 9.2. on a plane showing semantic density (OSD and DSD). There are three different patterns of reasoning evident. The bikeshare description (U1A) shows the design of the overall bikeshare scheme: bicycle selection, storage and exchange, the position of bike shelters for storage with consideration of access for maintenance, and the prescription of required safety gear. The bikeshare route (U1B) shows the disciplinary component of the design, the survey and generation of the DEM of the proposed modifications to the route taking into account the bikeshare description (U1A), users and terrain. The parking structure (S1) shows the loading analysis on the structure.

Discussion of the effects of recontextualizing choices in the design brief

The basis of recontextualization of the brief in U1 (Bikeshare scheme) is the weakening of DSD: both context and artifact can be adequately understood without requiring any specialized disciplinary knowledge. On the other hand, students need

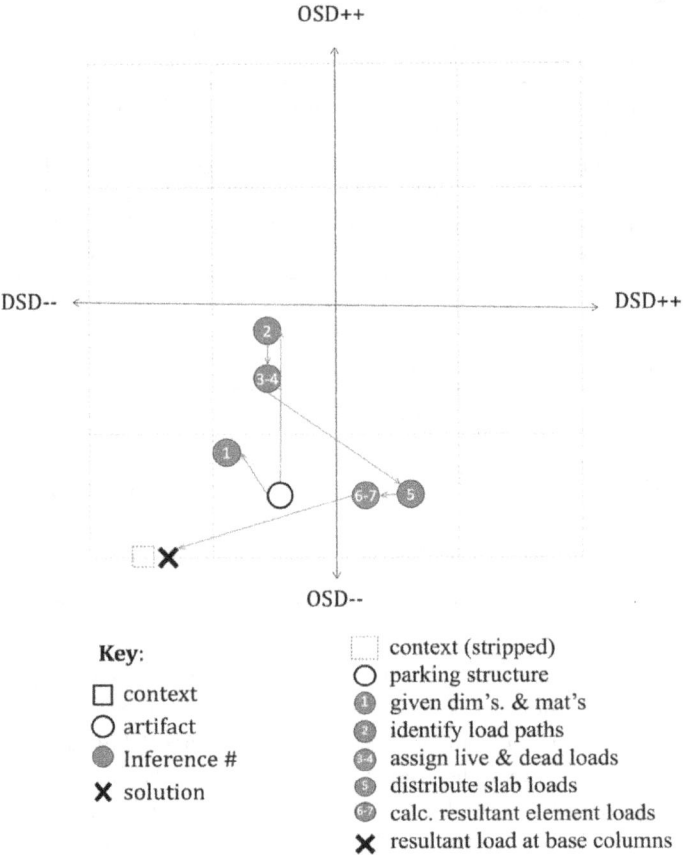

Key:

☐ context	⬜ context (stripped)
○ artifact	○ parking structure
⚫ Inference #	① given dim's. & mat's
✗ solution	② identify load paths
	③⁻⁴ assign live & dead loads
	⑤ distribute slab loads
	⑥⁻⁷ calc. resultant element loads
	✗ resultant load at base columns

FIGURE 9.4 Inferential reasoning: S1 (Parking structure)

to construct considerable understanding of the context and artifact to begin the design task: the strength of OSD is retained. By comparison, the basis of recontextualization of the brief in S1 (Parking structure) is the weakening of ontic relations OSD: both context and artifact can be adequately understood without engaging meaningfully with the context or the functionality of the structure. Here, the structure is constructed such that students need to engage with a number of different analytical approaches defined within the specialization of structural engineering: the strength of discursive relations DSD is retained.

Both principles of recontextualization have consequences. In the case of the bikeshare description (U1A), students are able to develop a solution based solely on common sense knowledge and experience without recourse to disciplinary knowledge. The science of logistics of the scheme, the structural analysis of the associated storage structures, formal principles of usage analysis were not required in the recontextualized project. Projects of this nature do *not* help students to learn to reason using specialized disciplinary knowledge in relation to the complex reality of the world. Instead students are able to develop an adequate solution based solely on common sense. However, by locating the project in a surveying course and providing a requirement to survey the route and propose modifications in terms of gradient changes and vertical elevation points for the bikeshare route design (U1B), disciplinary knowledge was introduced into the project and students were required to engage simultaneously with both with the complexity of the world (stronger OSD) and with disciplinary knowledge (stronger DSD).

In contrast, the recontextualized brief presented in S1 forces students to engage with structural engineering principles and procedures (stronger DSD), but in this case without any significant engagement with the complexity of the world (weaker OSD). Again, projects of this nature do *not* help students to learn to reason using specialized disciplinary knowledge in relation to the complex reality of the world. In this case students are able to develop an adequate solution based solely on disciplinary knowledge and procedures, without engaging with the complexity of a real object functioning in context.

The analysis of *inferential relations* that was described above was determined by the nature of the relations between concepts through the process of design. The details of the analysis of the inferential relations is elaborated in the section that follows. At this point suffice to say that the categorization of the inferential relations shown in Figure 9.2 is based on the pattern of reasoning between context, artifact and each step in the process of design from brief to final solution. In the case of bikeshare route design (U1B) the strength of the ontic relations of the artifact and context were retained, and certain discursive relations were required. Each step in the prescribed sequence of tasks then drew interdependently on the artifact, context and prior steps. In the case of the parking structure loading analysis (S1), the ontic relations were stripped from the context and weakened for the artifact. Although the discursive relations remained strong, the prescribed inferential steps became effectively independent of artifact and context, resulting in a linear sequence of reasoning.

In summary, weakening the semantic density of ontic relations risks dislocating the reasoning from the complexity of the world, resulting in a linear sequence of reasoning. Retaining the complexity of ontic relations without introducing a disciplinary knowledge component risks severing the links to professional knowledge. This is particularly a risk early in the curriculum when students have limited proficiency with specialized knowledge. In order to develop the skills needed to reason between the world (ontic relations) and specialized disciplinary knowledge (discursive relations), it is critical to retain or introduce both ontic and discursive components in the requirements of the design project. The design of the bikeshare route (U1B) provides one example of how this might be done.

The effect of recontextualizing choices on inferential reasoning

This section elaborates the categorization of the inferential relations presented above. The analysis is based on a list of sequential steps provided in an addendum to each brief. Each step was analyzed in terms of its own relative complexity and the relations to other steps in the process. Figures 9.3 and 9.4 show a comparison between the steps involved in the design of the bikeshare route (U1B) and the analysis of the parking structure (S1), respectively. What is immediately evident is the difference in the patterns of reasoning. Figure 9.3 (U1B) shows a network of inferences required in the development of a solution. Figure 9.4 (S1) shows a chain of inferences.

S1 (Parking structure), shows how weakening OSD to the point of dislocation resulted in a single input point (O) at the start of the chain of reasoning. With the context (□) stripped, it plays no role in the reasoning. What results is a linear sequence of potentially procedural calculations as the load is calculated at each level. With the retention of stronger OSD in U1B (Bikeshare route), both the artifact (O) and the context (□) are relevant to the development of the solution (✗). The interdependence between steps is retained and the resultant inferences resemble a network rather than a chain. For example, Step 4 in U1B requires proposals for modifying the terrain of the route. This draws on the interaction between artifact and context: the analysis of the campus (Step 1), the survey of the selected route (Step 2), and issues emergent from the description of the bikeshare scheme (Step 3). Decisions such as modification of the maximum operational gradient for inexpert/casual cyclists or the introduction of wider lanes to accommodate multimodal transport rely on a network of interdependent reasoning.

A comparison of the analysis of each project suggests that when ontic relations are stripped – as in S1 (Parking structure) – the resulting reasoning tends to be linear. On the other hand, despite stripping the discursive relations in the brief for U1 (Bikeshare scheme), the retention of the complexity of the ontic relations resulted in a more complex network of reasoning. However, without prescribing requirements for using disciplinary knowledge – surveying knowledge and procedures in the case of U1B (Bikeshare route) – the risk is that the semantic density of discursive relations will remain weak, as illustrated in the bikeshare description (U1A).

The influence of semantic gravity

One final point to make refers to the relationship between the semantic density of ontic relations (OSD) and semantic gravity (SG). The complexity of the ontic relations in U1 (Bikeshare scheme) is retained (OSD+), while in S1 (Parking structure) the ontic relations have been simplified to the point of dislocation (OSD– –) or striped of relevance completely (OSD_o). This tends to correspond with the strength of semantic gravity. The specificity of a context or artifact (SG↑) depends on the level of detail of ontic relations (OSD↑). On the other hand, as ontic detail is stripped (OSD↓) the object becomes more generic (SG↓). This suggests that retaining stronger OSD requires the brief to refer to more specific contexts and artifacts. But that leaves the question of redundancy of OSD. If the strength of OSD corresponds with the strength of SG, is it not redundant to introduce OSD? With reference to the analysis of the *inferential reasoning* in U1B (Bikeshare route) presented in Figure 9.3, the diagram shows a wide range of OSD, however the reasoning always relates to the specific solution of a specific campus (SG++). While there is some correspondence between OSD and SG, it is not inevitable. Even if a specific object is referenced (SG++) that object may be more or less complex in its own right (OSD can vary between OSD++ through OSD– – through a design). The context (university campus) referred to in the case of U1 (Bikeshare scheme) is extremely complex (OSD++), but the analysis of a particular slope on the selected route is significantly simplified (OSD–). Both relate to a specific context or part of the context (SG++).

Concluding remarks

Real artifacts and real contexts are extremely complex in their contextually embedded state. Making sense of them and simplifying them in order to make design decisions and to predict performance requires making sense of the contextual details, both in terms of unspecialized 'everyday' familiarity, and informed by disciplinary insights. Projects or 'problems' encountered in the 'real world' are specific, contextually embedded and emergent. In their emergent form they would typically be coded SG++, OSD++, DSD_0. While many 'problems' encountered in the world can be resolved without recourse to specialized knowledge, those projects or problems that fall into the preserve of any particular profession do assume the recruitment of the specialized knowledge associated with that profession. Those projects require the professional to strengthen the semantic density of discursive relations significantly (DSD↑) as they identify appropriate theoretical principles from the canon of professional knowledge (DSD++). In order to theoretically model any potential solution, the material complexity of the emergent problem (ontic relations) needs to be simplified; a process of weakening the semantic density of ontic relations (OSD↓) by identifying relevant aspects of the problem and discarding those that are not relevant. But this process is not independent of discursive relations; OSD↓ is likely to be informed by introducing principled insights (discursive relations), which involves DSD↑.

Students learning to become engineers or other professionals need to learn to identify and recruit appropriate specialized disciplinary knowledge to solve professional problems. Students should not be expected to develop this expertise without an explicit introduction, at an appropriate level of complexity. This is especially so early in the curriculum before students become proficient in multiple disciplines, or even have much familiarity with the objects of their profession (what they are and how they work). It is therefore necessary to simplify those projects intended to mimic professional projects to an appropriate level of complexity for any particular point in a learning trajectory. The two projects I have contrasted in this chapter appear at the start of the design learning trajectory and so required significant recontextualization. The analysis shows the way in which the *artifact prescription* and *context description* are recontextualized in the design brief affects the relationship between 'concrete' object and 'abstract' theory, modifying the inherently dialectical relation between them.

The argument I am making is that when the semantic density of ontic relations (OSD) remains stronger, the required reasoning is more complex. It is likely to result in a simultaneous network of inferences between ontic relations and discursive relations rather than a sequential chain of inferences. Of course, if the discursive relations are not relevant, the reasoning is also simplified, but in a different way, the reasoning never requires the semantic density of discursive relations (DSD) to be strengthened. The suggestion is that professional reasoning requires dialectical reasoning between ontic relations and discursive relations. If ontic relations are stripped then the reasoning becomes more procedural, while if discursive relations are stripped the range of semantic density is reduced, and the significance of specialized disciplinary knowledge is compromised.

I have not suggested any prescription of how one 'should' simplify design projects; there are many ways to do that. But if one is aware of the potential consequences of the recontextualization choices made, then one can be more intentional about designing coherent learning trajectories. In typical engineering curricula, based on learning decontextualized sciences, there is a tendency to weaken OSD to the point of dislocation. Unless one takes seriously the complexity inherent in objects in the world and the effect that they have on the complexity of reasoning required to analyze them we will continue to be surprised when students cannot 'just' apply theory to the sorts of complex, contextually emergent problems that professionals encounter in practice.

In addition to insights into the empirical challenge of educating professionals, this study offers insights for building a better model of professional knowledge. Models of professions based on what Bernstein (2000) called 'regions' argue that specialized knowledge should be taught first and can then be 'applied' to external problems. But regions face both internally to specialized bodies of knowledge and externally to their application in fields of professional practice. This suggests that both internal relations of coherence (conceptual – discursive relations) and external relations of coherence (contextual – ontic relations) need to be held simultaneously. There is a dialectical relationship between the internally defined disciplinary knowledge and the externally emergent problem in the world. It is thus not merely an application of specialized knowledge onto an emergent problem. Professional reasoning shifts continuously between the internal specialized knowledge *and* the external concerns of the problem.

This study contributes to understanding 'regions', an underdeveloped concept in code theory, and shows how LCT Semantics can provide a lens into the external objects of analysis. Moreover, the categorization framework offers an analytical language that may be useful to other professional education research and development studies. As education moves in the direction of regionalization and education moves towards a focus on application of knowledge in the world, the complexity of the world needs to be taken into account. Otherwise we run the risk of sliding into issues of personal attributes and generic skills. We risk losing the power of specialized disciplinary knowledge because very few students learn to use specialized knowledge in the contextually complex situations that are the workplace.

Note

1 For an earlier version of the analytical distinction between ontic relations and discursive relations, see Wolmarans (2014).

Reference

Beck, J (2002) 'The sacred and the profane in recent struggles to promote official pedagogic identities', *British Journal of Sociology of Education*, 23(4): 617–26.

Beck, J. and Young, M. (2005) 'The assault on the professions and the restructuring of academic and professional identities: A Bernsteinian analysis', *British Journal of Sociology of Education*, 26(2): 183–97.

Bernstein, B. (2000) *Pedagogy, Symbolic Control, and Identity: Theory, Research, Critique*, Oxford: Rowman and Littlefield.

Blackie, M. A. (2014) 'Creating semantic waves: Using Legitimation Code Theory as a tool to aid the teaching of chemistry', *Chemistry Education Research and Practice*, 15(4): 462–69.

Froyd, J. E., Wankat, P. C., and Smith, K. A. (2012) 'Five major shifts in 100 years of engineering education', *IEEE,* 100: 1344–60.

Georgiou, H., Maton, K. and Sharma, M. (2014) 'Recovering knowledge for science education research: Exploring the 'Icarus effect' in student work', *Canadian Journal of Science, Mathematics, and Technology Education*, 14(3): 252–68.

Grinter, L. E. (1955) 'Report on evaluation of engineering education', *Journal of Engineering Education*, 46(1): 25–63.

Harris, E. M. D. L., Grogan, W. R., Peden, I. C., and Whinnery, J. R. (1994) 'Journal of Engineering Education round table: Reflections on the Grinter report', *Journal of Engineering Education*, 83(1): 69–94.

King, J. (2007) *Educating Engineers for the 21st Century*, London: Royal Academy of Engineering.

King, R. (2008) *Engineers for the Future: Addressing the Supply and Quality of Australian engineering Graduates for the 21st Century*, Epping, NSW: Australian Council of Engineering Deans.

Kotta, L. (2011) '*Structural Conditioning and Mediation by Student Agency: A Case Study of Success in Chemical Engineering Design*', unpublished PhD thesis, University of Cape Town.

Mann, C. R. (1918) 'A study of engineering education', *Bulletin*, 11.

Maton, K. (2013) 'Making semantic waves: A key to cumulative knowledge-building', *Linguistics and Education*, 24(1): 8–22.

Maton, K. (2014) *Knowledge and Knowers: Towards a Realist Sociology of Education*, London: Routledge.

Maton, K. (2020) 'Semantic waves: Context, complexity and academic discourse', in J. R. Martin, K. Maton and Y. J. Doran (eds) *Accessing Academic Discourse: Systemic Functional Linguistics and Legitimation Code Theory*, London, Routledge, 59–85.

Shay, S., and Steyn, D. (2016) 'Enabling knowledge progression in vocational curricula: Design as a case study', in K. Maton, S. Hood, and S. Shay (eds) *Knowledge-Building: Educational Studies in Legitimation Code Theory*, London: Routledge, 138–57.

Smit, R. (2017) '*The Nature of Engineering and Science Knowledge in Curriculum: A Case Study in Thermodynamics*', unpublished PhD thesis, University of Cape Town.

Wolmarans, N. (2014) '*Exploring the role of disciplinary knowledge in engineering when learning to design*', paper presented at *Design Thinking Research Symposium*, West Lafayette, IN, USA.

Wolmarans, N. (2017a) 'The nature of professional reasoning: An analysis of design in the engineering curriculum', unpublished PhD thesis, University of Cape Town.

Wolmarans, N (2017b) '*Flexible curricula: Addressing the transition from engineering science to engineering design*', paper presented at *SASEE 2017*, Cape Town, South Africa, 14–15 June, pp 349–57.

10

GROUNDED LEARNING

Telling and showing in the language and paralanguage of a science lecture

Susan Hood and Jing Hao

Introduction

The case study presented in this chapter is one of a series which draws on systemic functional linguistics (SFL) and more broadly systemic functional semiotics (SFS) to explore the potential of different semiotic modes to support the building of disciplinary knowledge and values in live lectures (Hood 2020; Hao and Hood 2019; Hood and Lander 2016). The object of study in this instance is the teaching of molecular chemistry in a face-to-face undergraduate lecture. As is typical of this pedagogic context, interaction is mediated across multiple semiotic modes including spoken and written language and visual images (static and dynamic) on presentation slides. In addition, there is the often-overlooked semiotic potential of the lecturer's embodied meaning-making. Student access to this latter semiotic mode is largely restricted to participation in live lectures. This is certainly so to the extent that the lecturer's body language is a feature of interaction with visible co-present others (Wilkin and Holler 2011: 294). However, to date this mode of pedagogic interaction has received relatively little research attention. This study aims to explore how technical scientific knowledge is built in the intersemiosis of the oral mode of telling and the visual embodied mode of showing. We pursue this work with a goal to inform pedagogic policies and practices for supporting students as they move into new uncommon-sense understandings of their disciplinary fields, particularly in environments which increasingly promote online learning.

Functional perspectives on the discourses of science

Current attention to discourses of science, especially in educational contexts, builds on a long history in SFL. Notable early contributions include a trilogy of titles foregrounding modes of interaction: Lemke's *Talking Science* (1990), Halliday and Martin's *Writing Science* (1993) and Martin and Veel's *Reading Science* (1998). Significant recent additions have taken a more subject-specific approach, privileging variation within the generalized field of science (e.g. Doran 2018 on the language of mathematics in physics education; Hao 2020a on the discourse of undergraduate biology; see also Chapters 6 and 7 this volume). At the same time these recent works have extended theoretical modelling of knowledge from an SFL perspective, with contributions informing this study. There has also been a refocusing of earlier interest in modes of interaction, expanding research beyond spoken and written texts to consider the interplay with other kinds of semiosis. In discourses of science these include images in textbooks (e.g. Ge et al. 2018), animations (He and van Leeuwen 2020) and embodied expressions of value (Hao and Hood 2019). Here we continue to explore the cooperation of spoken language and body language, complementing work to date with a focus on the intersemiotic construal of technicality in the context of a lecture on molecular chemistry.

Stratification and metafunction in language

Intrinsic to an SFL approach is firstly a relational theory of meaning. In other words, verbal or visual signs are *not* understood as symbolizing a reality 'outside' of semiosis but as bringing 'reality' into being – as making meaning (Halliday 1978). Following Saussure (1974), meaning is understood as **valeur**, as emergent in the relationship which holds between any two choices in a semiotic system of choices.

Secondly SFL's concept of meaning is metafunctional. Ideational meaning construes a world around us and inside us. Interpersonal meaning enacts personal and social relationships with others. Textual meaning has a composing role, organizing other metafunctional choices as a coherent flow of meaning and assigning relative prominence (Halliday 1994; Halliday and Matthiessen 2014). This metafunctional perspective on language is complemented by SFL's modelling of language as a tri-stratal system of systems – namely, **discourse-semantic** systems focusing on discourse structure, **lexicogrammatical** systems focusing on clause, group/phrase and word structure, and **phonological** and **graphological** systems of expression (Martin 1992; Martin and Rose 2007). The intrinsic metafunctionality of language across strata correlates with the extrinsic functionality of context as register; the register variables of field, tenor and mode correlate respectively with ideational, interpersonal and textual meanings in language. These aspects of the theory are captured in Figure 10.1.

Our focus on the building of disciplinary knowledge in molecular chemistry orients this study primarily towards ideational meaning. From a tri-stratal perspective on language (discourse semantics – lexicogrammar – phonology/graphology), we approach the question of convergence of meanings in speech and embodied expression at the stratum of discourse semantics (Figure 10.1). This offers a level of abstraction, a "view from above" (Matthiessen 2007: 2) that can account for co-expression across different semiotic systems. Here we focus on intersemiotic relations in spoken language and paralanguage only, space preventing inclusion of for

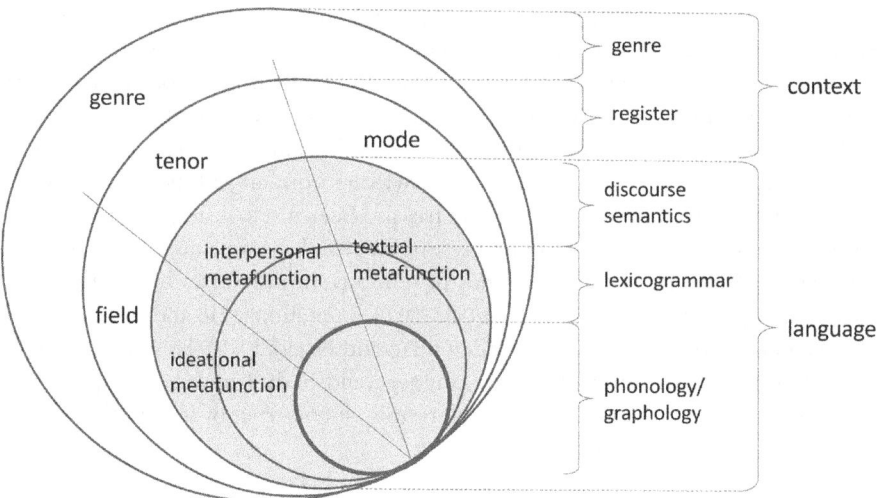

FIGURE 10.1 Stratification and metafunction in systemic functional linguistics

example visual and written verbal semiosis in presentation slides. We analyze first the spoken discourse of the lecture, then explore concurrent embodied expressions with reference to choices in the discourse system of IDEATION.

Before proceeding to more detailed discussion of ideational meaning-making in spoken language and the body, we introduce briefly a systemic functional modelling of body language as paralanguage.

Body language as paralanguage

As with language a social semiotic perspective on body language is metafunctional (e.g. Martinec 2000, 2001, 2002; Cléirigh 2011; Zappavigna et al., 2010; Hood 2011; Martin and Zappavigna 2019; Ngo *et al.* 2021). A brief account of the modelling of the semiotic potential of the body follows, with an explanation of why it is henceforth referred to as **paralanguage** in this chapter.

Setting aside protolinguistic expression (discussed in Ngo *et al.* 2021), SFL studies of paralanguage have identified two distinct systems (Cléirigh 2011; Martin and Zappavigna 2019). The first is one in which speech must be co-present for body movements to be meaningful. Here a full interpretation of textual prominence involves the synchronicity of body movements with the phonological rhythm and stress of speech (Halliday and Greaves 2008; Ngo *et al.* 2021) and a full interpretation of interpersonal tone involves body movements (e.g. of eyebrows, hands) corresponding with pitch movement (e.g. rising/falling). Cléirigh (2011) refers to this system as 'linguistic' body language. Following Martin and Zappavigna (2019) we refer to these systems as **sonovergent paralanguage**, using terminology which makes explicit the necessity of accompanying speech. Sonovergent paralanguage can express textual or interpersonal (but not ideational) meaning (Martin and

Zappavigna 2019). For ideational meaning we need to explore a second system, one which can occur with or without immediately co-present (convergent) speech. Cléirigh (2011) refers to this as epilinguistic body language. We refer to these resources as **semovergent paralanguage** (Hao and Hood 2019; Martin and Zappavigna 2019; Ngo *et al.* 2021).

Semovergent paralanguage relates meaningfully to discourse semantics across all three metafunctions (Martin 1992; Martin and Rose 2007). Interpersonally, it can realize options in the system of APPRAISAL – through facial expression of ATTITUDE as affect, through tension or size of gestures expressing GRADUATION, and through postures and positionings of the body expressing expansion or contraction of space for negotiating values in ENGAGEMENT (Painter et al. 2013; Ngo *et al.* 2021; Hao and Hood 2019). Textually, embodied paralanguage can realize choices in IDENTIFICATION, expressed in kinds of pointing gestures; it also converges with choices in PERIODICITY through longer wavelengths of whole-body movement and positioning in space (Martinec 2002; Hood 2011; Ngo *et al.* 2021). In this chapter we focus on the potential for expressing ideational meanings concurrently in speech and paralanguage.[1]

The systems of sonovergence and semovergence are both theorized as para- to language in that whenever speech is co-present there is synchrony with the prosodic phonology and in periods of absent speech, the rhythm of prior speech is typically maintained. Furthermore when speech is not immediately co-present, embodied expressions are interpreted as meaningful in ways dependent on the flow of spoken discourse (Martin and Zappavigna 2019; Hao and Hood 2019; Ngo *et al.* 2021).

The study

Our data come from a video recording of a 50-minute live lecture in molecular chemistry to a class of about 180 undergraduate students. Multiple semiotic modes are deployed and interact to different degrees as the discourse unfolds. Throughout, the lecturer deploys presentation slides displaying written language, static images and video clips. The slides are projected onto two large screens set behind and above the head of the lecturer. The lecturer takes up a position around a lectern in a center space at the front of the room. From there he faces the general student body, while periodically re-orienting himself to identify aspects of visual or verbal information on slides or to interact with individual students. The lecturer's spoken language and any audible student talk were transcribed and reviewed along with corresponding slides to identify the textual structuring of the content. Phases of discourse which share a similar function in building the scientific field and in which paralinguistic expressions accompany language were identified. The process modelled in this chapter is first to analyze linguistic expressions of scientific knowledge with reference to the discourse system of IDEATION and then to explore patterns of convergence with paralanguage. Both to foreground the contribution of the body and for reasons of length we limit the scope to the modalities of spoken language and paralanguage, while acknowledging the relevance of written text and images on slides in the multimodal discourse of knowledge building.

Building the field of molecular chemistry in language and paralanguage

Interest in how the field of molecular chemistry is built intersemiotically in a live lecture privileges an ideational perspective and explores choices in the discourse system of IDEATION. To contextualize this analysis, we first take account of the meaning potential at the higher stratum of register – focusing on field (Figure 10.1). In a pedagogic context such as a live lecture multiple fields can be involved in the discourse. One is the field of pedagogy which organizes classroom interactions and planning. This is realized for example through the underlined wording in *Let's talk about solid chemistry*, or in the lecturer's relatively frequent use of personal pronouns like *we/us/our*. At times pedagogic discourse is drawn on to present disciplinary knowledge of molecular chemistry in more commonsense terms as 'doing', as in *we can also get covalent network solids*, or 'having' as in *we have solids, liquids and gases*. Where helpful in analyses we have added bracketed paraphrasing which reinterprets from pedagogic to scientific discourse, as in *we can also get (there are also…)*.

Setting aside the pedagogic discourse, we tease apart two specific topics involved in disciplinary knowledge building in this lecture. The first is that of established (*observed*) knowledge. In our data this includes taxonomies and activities in the field of molecular chemistry. The second is that of experimentation, the (observing) procedures which underpin established knowledge (Hao 2020a; Hood 2010). Each of these topics can be explored from two simultaneous orientations, one static and the other dynamic (Doran and Martin this volume). A static perspective models field as **items**, their **properties**, and their organization into taxonomies. A dynamic perspective models field as **activities**, their **properties** and how activities develop as time unfolds. Specific sciences are distinguished by distinctive types of items, activities and properties and the relationships which hold between them (Martin 1992, Hao 2020a, this volume, Doran and Martin this volume).

We elaborate briefly on the basic discourse semantic units of IDEATION which realize field. From a static perspective, relevant units include **entities** (e.g. *molecules, atoms, sodium chloride*), **qualities** which extend those entities (e.g. *sodium is smaller*) and **dimensions** which are dependent on those entities (e.g. *a type of solid*).[2] Entities, their qualities and dimensions are presented and related in discourse in what are referred to as **state figures** (e.g. *sodium is smaller; one type of solid is ionic solid*). From a dynamic perspective the basic discourse units are **occurrences** (e.g. *the lattice expands*), **occurrence figures** which configure occurrences and entities (e.g. *one chlorine ion is attracted to one sodium ion*), and finally **sequences** of figures (e.g. *as our solution is being formed … our solvent begins to pull apart our solid particles, and surround them*). These units of ideational meaning are introduced and explained in turn with reference to a number of indicative phases of the lecture's talk.

From analyses of IDEATION in spoken language we explore convergent expressions in the lecturer's embodied paralanguage. We consider relations in meaning between what is told and what is shown through embodied expressions. The theorizing of meaning-making in intersemiotic convergences continues to present a considerable challenge to the broad field of multimodality (Zhao 2010; Martin 2011). The approach taken here applies the concepts of **instantiation** and **commitment** introduced below (and explained in more depth in Martin (2011),Painter, *et al.* (2013)).

Instantiation refers to a hierarchy of generalization in a system of meaning. The upper point of the cline embraces the full meaning potential of a system – a potential progressively constrained by selections of genre and register, through types of texts to actual instances of discourse (Martin 2008, 2011). This concept is relevant to explorations of how much meaning potential is **committed** from the system in instances of discourse (Martin 2008; Hood 2008).

Commitment offers a framework for comparing instances within and across semiotic modes. In this chapter we examine the relative commitment of ideational meaning in language and paralanguage by considering the degree to which activities and items of a given field construe a technical reality or an everyday commonsense one. Technicality condenses a considerable mass of ideational meaning through relations to other phenomena, within a given text and in the field generally (Martin 2017). Much more ideational meaning potential is committed in naming technical phenomena than in naming everyday commonsense things and events. This variation strongly differentiates the two modalities of language and paralanguage in the data. The verbal discourse of the disciplinary field is highly technical while the paralanguage expresses an everyday commonsense ideational reality, as is made evident in comparative instances throughout the chapter.

Variations in how ideational meanings are instantiated in discourse can also be considered from the perspective of textual meaning and mode. The focus then is on the role which language plays in an interaction, with language accompanying action at one end of a continuum and language as reflection at the other. These variations are associated with 'more written-like' or 'more spoken-like' discourse (Martin 1992; Martin and Rose 2007).

In comparing the relative commitment of ideational meaning in instances of language and paralanguage, we also consider the mode of expression as relatively **congruent** or **experientially metaphorical**. This contrast is exemplified below for the discourse semantic systems of IDEATION and CONNEXION (Martin and Matruglio 2020; Hao 2018, 2020a; Hood 2020).

For IDEATION, the expressions in focus are underlined below.

(we) <u>added</u> more and more sodium acetate
vs.
<u>the addition</u> of one single sodium acetate crystal…

The first instance (*added*) is a congruent realization of Process in a material clause. The second (*addition*) is experientially metaphorical in that it construes two layers of meaning – a Participant in the clause and an occurrence figure in the discourse semantics.

For CONNEXION, a contrast between a congruent and a logically metaphorical expression of a causal connexion is exemplified in:

> *Water molecules huddle around the ions <u>and so</u> the lattice begins to break apart*
> vs.
> *The action of the water molecules <u>leads to</u> the break-up of the lattice.*

In the first, the causal connexion between two figures is construed congruently through a conjunction. In the second, the causal connexion is construed metaphorically through a grammatical process.

While language allows us to shift between more or less congruent and metaphorical construals of an ideational reality – variations we associate with 'more written-like' or 'more spoken-like'– paralanguage does not. We can only gesture an entity metaphorically construed in languages as if it were a congruent entity. In other words, paralanguage always construes a 'congruent' construal of an ideational 'reality' – one grounded in a here-and-now. In this it may be convergent or divergent with accompanying verbal text.[3]

In the sections to follow we progressively explore the lecturer's spoken language and concurrent paralanguage in terms of the relative congruence that they instantiate. Several indicative extracts of lecturer talk are analyzed for choices in specific units of IDEATION. Concurrent paralinguistic expressions are then described and interpreted. We consider how the two modalities collaborate in the construal of ideational meaning, and how this supports students as they move into new uncommon-sense understandings of their disciplinary fields.

Entities and their relation in discourse

The IDEATION unit **entity** is differentiated into **thing entities** naming items (such as *molecules*, *atoms*), **activity entities** naming activities (the technical naming of a sequence of occurrences as in *crystallization*) and **semiotic entities** naming meaning (e.g. *an example*; *the periodic table*) (Hao this volume). The first two types are of particular relevance in this study. We begin with a focus on thing entities, their role in a static perspective on field knowledge, and their convergence with paralinguistic expressions. Activity entities and concurrent paralinguistic expressions are explored later where a dynamic perspective is adopted.

The first episode of talk for analysis of IDEATION comes at about the four-minute mark in the 50-minute lecture. It follows a brief review of the previous lecture and sets up the focus of a segment to come. This phase is densely packed with technical entities. In the notation system deployed in text (A) and other phases of analysis, entities are shown in bold. As our focus is on disciplinary knowledge, the discourse of the pedagogic field is bracketed off – as in [*But now let's talk about*]. In occasional superscripts we insert meanings elided in talk but recoverable from the co-text. As will become evident, this annotation supports the identification of relations which hold between entities in the phase of text. In round brackets we offer some paraphrasing from pedagogic into scientific discourse and substitutes of highly generalized terms such as *something* and *thing*, as in *so something (/examples) like solid sulphur)*.

---

text (A)

[But now let's talk about] **solid chemistry**, [just for a little bit]. [So] even though **solid** is just a state of **matter** – [so we have] (there are) **solids, liquids**, and **gases** – the internal structure of [our] **solids** can be different depending on what **atoms** [we have] (are) present. [So] there are a few different types ^{of solid} [and we're going to go through them in the next few slides]. The one ^{type of solid} [we're all fairly acquainted with at this point] is **ionic solids**. [So] the main example ^{of ionic solid} is **sodium chloride**. [We can also get] (there are also) **molecular solids**, so something (/examples) like **solid sulphur**. [We can also get] (there are also) **covalent network solids**, so these are things (/examples) like **diamond, sand, graphite**. **Metallic solids** is [our] (the) last one ^{type of solid}.

As in any discourse semantic system, semantic choices may be realized in more than one kind of grammatical construction (Hao this volume). Our analysis of language focuses primarily on choices at the discourse semantic stratum.

Entities in paralanguage

While the grammatical realization of entities in discourse varies, all except one of those bolded in text (A) *(atom)* are concurrently expressed in paralanguage; and for all of these the embodied entity is sculpted through the shaping of the palms of both hands, oriented to each other to some degree as in Figure 10.2.

Figure 10.2 illustrates a default expression of a sculpted paralinguistic entity in our data – i.e. one which generally requires minimal effort to express, with relaxed curved hands, palms oriented to each other, with forearms extended out from the body from waist to chest height and at torso width. Variations on this shape, size and position commonly associate with the depiction of particular qualities of an entity.

In some cases an entity is expressed through a single-hand shape – as in the upwardly (supine) curved palm in Figure 10.3, and later, in Figure 10.12, an inverted (prone) palm. We hereafter refer to these as **paralinguistic entities**.

Elsewhere in the data paralinguistic entities express more defined shapes, typically requiring two hands. Concurrent with the entity *box* in (1) is a paralinguistic entity in which straightened hands more sharply define the boundaries, as in Figure 10.4. Relative to the default expression (Figure 10.2), the paralinguistic entity in Figure 10.4 commits more meaning potential – here associated with the straight-sided quality of the entity *box*.

FIGURE 10.2 Two hands sculpting a paralinguistic entity

(1) it's a bit like taking a 1cm by 1cm <u>box</u> out of our ionic solid (...)

FIGURE 10.3 One hand sculpting a paralinguistic entity

Concurrent with the underlined dimension in (2) is a sculpted paralinguistic entity in which digits of both hands depict a kind of pyramid shape, as in Figure 10.5. In relation to Figure 10.2, the paralinguistic expression commits more meaning potential – here associated with the quality of a 3D entity.

FIGURE 10.4 A paralinguistic entity convergent with 'box'

(2) it (the inner lattice) [wants to] <u>form into</u> three dimensions

FIGURE 10.5 A paralinguistic entity convergent with 'form into three dimensions'

An alternative means for depicting a paralinguistic entity is to 'draw' an object in the air with one or more fingers or hands. Sculpted or drawn entities can vary in size, shape and location in the gestural space. A smaller entity for example can be sculpted in a single hand (Figure 10.3 above), and the depiction in Figure 10.2 might be expanded or reduced in size to commit more meaning of big-ness or small-ness. The depiction in Figure 10.3 might be duplicated with entities in both the left and right hand. This visual depiction might co-commit a meaning of two entities in concurrent verbal text. However, as this is the maximum number of entities able to be expressed (statically) in the two-handed body, the gesture can at times under-commit a meaning of *lots of* (i.e., more than two) in a verbal text.

With reference to the spoken text (A), the technical entities of the disciplinary field of molecular chemistry (*solids*; *atoms*; *sodium chloride*; etc.) each commit a considerable mass of ideational meaning, infused as each is with a complex set of relations to other technicality in the field (Martin 2017). The commitment of ideational meaning in these instances is considerably greater than in concurrent paralinguistic entities. At the same time the concurrent paralanguage construes the meaning of entities in commonsense terms, depicting each as a hand-held thing. This establishes an important complementarity of technical telling and everyday showing. It grounds the technicality in commonsense.

Relations between entities in language

The discourse of text (A) not only construes multiple technical thing entities, but also explicitly construes the ways entities relate to one another. A detailed set of relations among entities is established in this relatively short phase. The related entities in Figure 10.6 all construe relations of classification building from the general entity *matter*.

In text (A) each entity is meaningful not in isolation but in relation to its superordinate category, its co-classes and subclasses. All the entities in Figure 10.6 co-classify each other. For example, *matter* and *solid* co-classify as superordinate and subordinate respectively, and *ionic solids* and *molecular solids* co-classify as co-subtypes. The co-classification between entities can be grammaticalized in a number of ways: for example in a relational process, as in (3).

(3) **solid** is just a state of **matter**

Another possibility is a nominal group complex as in (4).

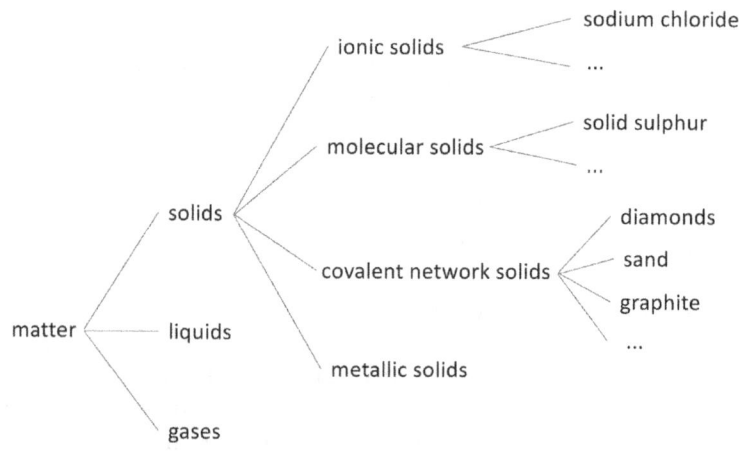

FIGURE 10.6 The taxonomy of *matter* configured in text (A)

(4) [We can also get] **molecular solids**, [so] something like solid **sulphur**

Additionally, relations between entities in the spoken discourse can be abduced (interpreted from the co-text), in some cases across extended stretches of text. For example, in text (A), *molecular solids* and *covalent network solids* are introduced as co-subclasses of the category of *solids*. The clues supporting this abduction include the distinguishing Classifiers of the *solids* (underlined) and the textual reiteration of *We can also get*, bracketed in (5).

(5) [We can also get] **molecular** solids ... [We can also get] **covalent** network solids ...

A further discourse semantic choice of IDEATION which supports the construal of taxonomic relations among entities is that of **dimensions** (Hao this volume). Dimensions explicitly name those taxonomic relations. The instances of dimension underlined below, *type* and *example*, explicitly name the relation between the entities involved as one of classification. The superscript notation shows that the relevant entities are elided, having been introduced in preceding text.

(6) [So] there are a few different types $^{\text{of solid}}$

(7) the main example $^{\text{of ionic solid}}$ is **sodium chloride**.

Relations between entities in paralanguage

A taxonomic relation of classification between entities in text (A) is almost always concurrently expressed paralinguistically through paralinguistic entities which depict relative size and/or shape and/or location. Several instances exemplify this.

Two kinds of classification relation are construed in the immediately sequential state figures in (8) and (9).

(8) even though **solid** is just a state of **matter** ...

(9) we have **solids, liquids**, and **gases**

In (8), there is the classification relation of the class *matter* with the subclass *solid*. In (9), the tshree entities *solids*, *liquids*, and *gases* are related to each other as co-subclasses to *matter*.

In (8), a class-subclass relation is made explicit in the dimension *state* of the entity *matter*. In convergent paralanguage, one entity is expressed concurrently with *solid* and another with *matter*. That convergent with *solid* differs in size, shape and location from that with *matter*, as described in Table 10.1.

Correspondences evident in Table 10.1 generalize to the data as a whole. When the verbal construal of a class-subclass classification relation converges with the expression of paralinguistic entities, they are depicted differently with respect to one or more of the features of size, shape and location. Non-equivalence on a taxonomic hierarchy converges with non-equivalence in depicted entities. In some instances size is a variant, but more consistently there is a variation in shape, in line

TABLE 10.1 Paralinguistic expressions of the relation of *solid* to *matter* as subclass to class

Language		
entity	*solid*	*matter*
relation	Subclass	Class
Paralanguage		
size	smaller than default (median)	larger than default (median)
shape	sculpted: palms facing; thumbs up; defined boundaries with straightened fingers	sculpted: more open curved palms; undefined boundaries
location	located at chest height	located below chest height

⇦···· *gases* ···························· *liquids* ···························· *solids* ······

FIGURE 10.7 Equivalence in the paralinguistic depiction of co-class entities in (9) – viewed from right to left

with that described in Table 10.1. The paralinguistic entity convergent with class is depicted as sculpted in soft curved palms (akin to Figure 10.2), and that convergent with subclass is depicted with more sharply defined borders (akin to Figure 10.4). This suggests that a paralinguistic entity with more defined boundaries can not only commit a specific quality of an entity (as having straight sides as in '*box*') but can more generally express a 'kind of' taxonomic relation.

In (9), the state figure (*we have* **solids**, **liquids**, *and* **gases**) presents a different classification relation. Here, the three entities are related to each other as co-subclasses of *matter*. The co-subclass relation establishes them as equivalent on a classification hierarchy. This equivalence is concurrently expressed in the depiction of three paralinguistic entities, each showing equivalence in size, shape and location (Figure 10.7 and Table 10.2).

The taxonomic relations between the entities in (9) (*solids, liquids and gases*) are committed differently in each semiotic mode. In the spoken text, their equal status in a taxonomic hierarchy has to be inferred from verbal co-text. The paralanguage is more explicit in its depiction of three entities equivalent in size, shape and location. In this instance a higher level of ideational commitment is construed in the showing than in the telling of the relations between the entities.

TABLE 10.2 Paralinguistic expressions of co-subclass relations among *solid, liquids* and *gases*

Language			
entity	*solids*	*liquids*	*gases*
relation	co-subclass	co-subclass	co-subclass
Paralanguage			
size	equivalent in size: median		
shape	equivalent in shape: sculpted in two hands; relatively defined sides		
location	equivalent in location: along same horizontal plane; in front of body just above waist height		
	[Note: torso rotates from the lecture's left to right with each paralinguistic entity. From the students' view the entities move from right to left in space with equal distance between them]		

A further instance of co-subclass relations is shown in (10).

(10) How do we go from (/distinguish) **a primitive** cubic to **a body**-centred cubic to **a face-centered cubic**?

The concurrent paralanguage shows a similar pattern of intersemiotic convergence to Figure 10.7. The bolded entities in (10) construe three co-subclasses of *cubic* (structure). Again, the co-relations are expressed through paralinguistic entities of equivalent size, with relatively defined boundaries, and located equidistant from one another along a horizontal plane (Figure 10.8).

In (9) and (10) the language construes the co-subclass entities as an exhaustive set of kinds of *matter* in (9) and kinds of *cubic* in (10). In each case the set of concurrent paralinguistic entities are depicted with more sharply defined boundaries, with relatively straightened palms.

In other instances of similar relations, the entities are construed as non-exhaustive sets, in other words as some of, or examples of a set of co-subclass entities, as in (11).

◄··· *a face-centred cubic* ············ *a body* ······················· *a primitive* ·····

FIGURE 10.8 Equivalence in the depiction of an exhaustive set of co-related paralinguistic entities (in 10) – viewed from right to left

(11) [We can also get] **covalent network solids,** [so] these are things like **diamond, sand, graphite**

In (11) an occurrence figure followed by a state figure construes *diamond, sand, graphite* as a set of co-subclass entities to the class *covalent network solids.* Their status as examples only is construed in the generalized reference to more such entities in *things like.* The non-exhaustive status of this set corresponds to a difference (from Figures 10.7 and 10.8) in the depiction of paralinguistic entities in Figure 10.9.

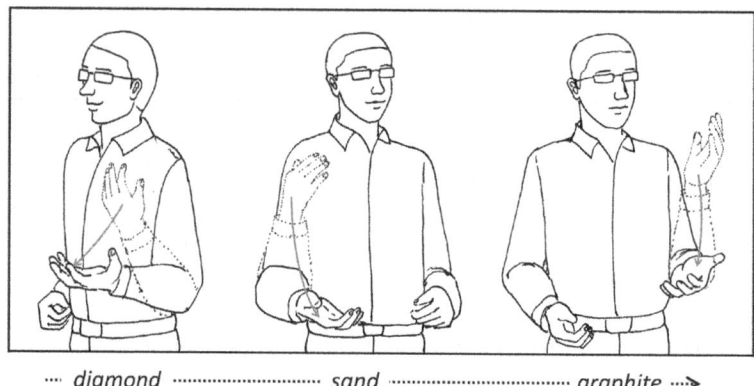

···· *diamond* ·························· *sand* ·························· *graphite* ···>

FIGURE 10.9 Equivalence in the depiction of a non-exhaustive set of co–subclass entities (in 11) – viewed from left to right

In contrast to the exhaustive sets of co-related entities depicted in Figures 10.7 and 10.8 as paralinguistic entities with relatively defined boundaries, the non-exhaustive set depicted in Figure 10.9 is sculpted in single hands, but more significantly in hands with an open, curved palm. There is also a difference in how the paralinguistic entities are located in relation to each other in the gestural space. In Figures 10.7 and 10.8 they are spaced out equidistant from one another along a horizontal plane. In Figure 10.9 they are located in the same general space in front of the body, depicted in alternating left, right then left hands.

The general pattern of convergence in (11) and Figure 10.9 is repeated in relation to (12). The text again construes a non-exhaustive set of co-related entities, with *pure* and *compound* construed as a non-exhaustive set of kinds of *solution.*

(12) [We spoke about] **solutions,** and **pure, compound,** all those <u>sorts of</u> **things,** [way back in the first lecture]

Their shared sub-class status of *pure* and *compound* is construed in the dimension *sorts of.* Their status as a non-exhaustive set is construed in the generalized reference to more entities of this kind in *all those sorts of things.*

Instance (12) is similar to (9), (10) and (11) in that the concurrent paralinguistic entities for *pure, compound,* and generalized others in *all those sorts of things* are depicted as equivalent in size, shape and relative location (Figure 10.10).

····· *pure* ································· *compound* ··················· *all those sorts* ····>
of things

FIGURE 10.10 Equivalence in the depiction of a non-exhaustive set of co-class entities (in 12) – viewed from left to right

Each paralinguistic entity in Figure 10.10 is expressed as more or less equivalent in shape and size, with the proviso that in the first (*'pure'*), the lecturer had lowered his left hand to notes on the desk. Here the right hand sculpts the paralinguistic entity palm down over the lowered left hand. The left hand is then freed to come into play to sculpt the other entities (*compound* and *those sorts of things*) as two-handed. Each entity is located more or less in the same space in front of the torso. A rotation in the angle of the two sculpting hands shifts the location of each para-linguistic entity a little to the right or left side of the torso.

The instances of intersemiotic convergence in Figures 10.7, 10.8, 10.9 and 10.10 all depict co-relations within sets of paralinguistic entities through equivalence in their depictions. However, where the set of co-related entities is exhaustive, the paralinguistic entities are sculpted in hands which depict more defined boundaries and are viewed as placed more or less equidistantly along a horizontal plane (Figures 10.7 and 10.8). Both features reinforce a stronger categorical perspective on the co-relations between entities concurrent with the verbal text. Where the set of co-related entities are non-exhaustive, the paralinguistic entities are sculpted with less defined boundaries in softer curved palms. They are viewed as located more or less in the same location in front of the body (Figures 10.9 and 10.10). Concurrent with the verbal text these features foreground commonality or likeness over categorical distinctions in entity relations.

Entities and related qualities

Qualities of entities in language

A focus on the qualities of entities shifts our perspective from relations of classification to those of composition. At about eight minutes into the lecture, after establishing the taxonomic relations in text (A), the lecturer narrows the focus to one type of solid – ionic solids, and discusses its delicate sub-classes based on various compositions. The critical knowledge to be established in this part of the lecture is not only a recognition of particular types of composition, but importantly an understanding of why they are different. To do this the lecturer describes qualities of the ions which compose them. This phase is presented as text (B), with entities in bold, qualities in bold and italics, and dimensions of entities underlined.

---

text (B)

[So] the <u>structure</u> of our **internal solids** are dependent on two things – the first one is <u>charge</u> ^{of relevant atoms} and the second one is <u>size</u> ^{of relevant atoms}. [So] let's look at <u>size</u> ^{of relevant atoms} first. [So] we've got **sodium chloride**, and **cesium chloride**. [So] **sodium** is *very small,* and **cesium** is *very big,* further down on our period table. These are what their <u>ionic structures</u> look like (*pointing momentarily to images on PPT slide*). We have the **sodium chloride** from the previous example. Each **sodium** has six **chlorines**, each **chlorine** has six **sodiums**. But because ^{the size of} **sodium** is *smaller* it (/**sodium chloride**)'s able to have that <u>six to one ratio</u>. ^{The size} ^{of} **Cesium** is *bigger.* So **cesium** *takes up more space* than **sodium**. So it (/**cesium**)'s going to come into contact with more **ions**. (...) [So] because it (**cesium**)'s in contact with more **ions**, its (**cesium chloride's**) <u>structure</u> is going to be different.

Text (B) begins by introducing two **dimensions** of entities, those of *size* and *charge*, with size the focus of the phase. Dimensions in text (A) mostly name classification relations (*type of; example of*). In text (B) the dimension of *size* names a measurable property of an item in field (Doran and Martin this volume). Properties of items are realized in discourse by qualities. In (13), qualities such as *very small, very big, smaller* and *bigger* extend the meanings of the entities *sodium* and *cesium*.

(13) ... ^{the <u>size</u> of} **sodium** is *smaller* (...) ^{The <u>size</u> of} **Cesium** is *bigger.*

The qualities in (13) are realized grammatically through Attributes in relational attributive clauses, in which the Head of the nominal group is an Epithet. Other grammatical realizations of qualities follow in later instances.

Qualities of entities in paralanguage

In spoken language a quality is always ascribed to an entity, either in the grammar or in the co-text. In paralanguage too, an entity is required for the depiction of a quality. The entity and the quality may be expressed as separate paralinguistic entities (one convergent with the entity and one with the quality). This is the case for the state figures in (14) and (15). Each underline indicates concurrence of a paralinguistic entity.

(14) sodium is smaller
(15) cesium is bigger

While not illustrated here, the quality *smaller* is convergent with a sculpted paralinguistic entity which is smaller in size than that convergent with *bigger*.

In other instances, state figures which construe an entity and a quality can be concurrently depicted in a single paralinguistic entity. This is the case for both (16) and (17).

(16) sodium is very small
(17) cesium is very big

The state figures in (16) and (17) occur sequentially in text (B). The concurrent paralanguage for the sequence of figures is shown in Figure 10.11.

Concurrent with (16), in the faded depiction on the right, a small paralinguistic entity is sculpted in a single hand. The lecturer's torso is oriented to his left. Concurrent with (17) a larger paralinguistic entity is sculpted with two hands. The lecturer's torso is oriented to his right. The meaning of relative small-ness or big-ness is co-committed in language and paralanguage. The contrasting locations of the paralinguistic entities as the torso rotates left then right additionally commit the contrastive relation of the qualities of small and big.

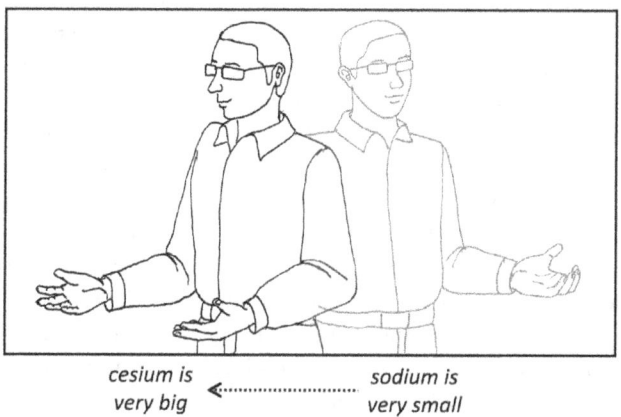

cesium is   *sodium is*
very big ←.................. *very small*

FIGURE 10.11   Paralinguistic entities concurrent sequentially with (16) and (17) – viewed from right to left

In another instance the quality *sticky* of the entity *adhesive force* is realized verbally through the grammatical Attribute bolded in (18).

(18) The adhesive force is the intermolecular forces that bind a substance to something else – how **sticky** is it to another surface.

Concurrent with '*how sticky is it*' a paralinguistic entity is expressed dynamically as an entity in motion – an 'entitied occurrence' in the discourse semantics as explained shortly (Figure 10.12).

FIGURE 10.12 A paralinguistic expression depicting a quality as an entitied occurrence concurrent with 'how sticky is it'

The paralinguistic entity in Figure 10.12 is sculpted in one-hand, with a prone (down-facing) palm and fingers extended. The occurrence is depicted dynamically in the lifting of the paralinguistic entity in a restrained motion, with the extended fingers coming together. In the flow of spoken language, the lecturer's construal of a technical field in *adhesive force* [as] *intermolecular forces that bind a substance to something else* is reinterpreted into a commonsense one in *how sticky is it*. The technicality of the former commits considerably more ideational meaning potential than does the commonsense of the latter. The paralanguage (Figure 10.12) co-commits the commonsense not the technical construal. The dynamic perspective on field introduced in this instance is the focus for the next section, but we'll briefly consolidate the static perspective on intersemiosis in building disciplinary knowledge taken to this point.

The spoken language of the lecture on molecular chemistry construes multiple technical entities, qualities which extend their meanings, and relations which hold between them. An understanding of all is required in appreciating the relevant established knowledge of the disciplinary field. The required technical load is considerable, both in terms of the mass of meaning condensed in individual entities (Martin 2017), and also in the density of technical entities in the discourse. We note the means by which this mass of technicality is complemented in paralanguage through the construal of commonsense entities, qualities and relations between entities.

While not discussed to this point, in all instances the paralinguistic showing of ideational meanings also construes them as congruent – as realizing phenomena in the here-and-now of a shared material space. The intersemiosis of spoken language and paralanguage in the lecturer's discourse bridges the technical and the everyday. It grounds meanings in a congruent material ideational reality in support of building knowledge of specialized items and properties in the field.

Dynamic construal of knowledge in language

Taking a dynamic perspective on disciplinary knowledge, the focus shifts to scientific activities and their unfolding through time. We begin by considering how activities in the field of molecular chemistry are realized in the discourse of spoken language and then in the intersemiosis with concurrent paralanguage.

Occurrences and occurrence figures in language

In language, activities are expressed in the ideational unit of an **occurrence figure**, which configures an **occurrence** and associated **entities**. Occurrence figures can also unfold in a sequence linked through logical relations. This ideational unit is explored with reference to a number of phases of the lecturer's talk.

Occurrence in language

At around 30 minutes into the lecture, having built knowledge of different types of solid, their qualities and internal structures, the lecturer begins a stage in which he explains an activity which relates to 'intermolecular forces' and the formation of a solution. The lecturer uses the example of salt dissolving in water, as presented in text (C). Here and elsewhere in this section occurrences are bolded and underlined and associated entities bolded without underlining.

---

text (C)

And as our **solution** is **being formed** – [so we think about] **dissolving** **salt** – our **solvent** begins to **pull apart** our **solid particles** and **surrounds** **them** (solid particles). [So our depiction of] **salt** **dissolving** [is we] start with our **lattice**. (...) Those are our **water molecules** **huddling around** our different **ions** to **break** **it** **(lattice)** **apart**. And **it** **(lattice)** will be **uniformly dispersed** throughout the **solution**.

Occurrences in text (C) are realized grammatically by Processes in material clauses. They represent generalized scientific phenomena – phenomena observable in different locations in time and space.

Occurrence figures in language and paralanguage

An **occurrence figure** is a larger ideational unit which construes an activity. It configures an occurrence with one or more associated entities. These entities may function in one of two roles: they may precipitate an occurrence, or they may be precipitated by one. In (19), the single entity *solution* is the precipitated. In (20), *solvent* is the precipitator of the occurrence *pull apart* and *solid particles* are the precipitated.

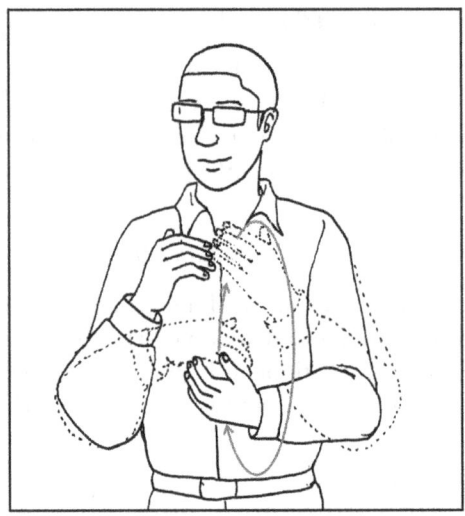

FIGURE 10.13   The paralinguistic occurrence convergent with 'being formed' (in 19)

(19) … as our solution is <u>being formed</u>
(20) our solvent begins to <u>pull apart</u> our solid particles.

Almost all instances of occurrence in text (C) are concurrently construed in paralanguage. The paralinguistic **occurrences** are realized through motion, typically expressed in a hand and forearm or arm. Potential variations include the trajectory of the motion through the gestural space (linear, curved or circular), the kind of flow depicted (relatively restrained or accelerated) and whether the movement is iterated or not (Ngo *et al.* 2021). For example, concurrent with the bold and underlined occurrence in (19), the lecturer rotates both hands and forearms around each other in a clockwise direction outwards from the body at chest height (Figure 10.13).

This circular, iterative, and relatively accelerated motion depicts an occurrence as unfolding in time. The gesture co-commits with the verbal *being formed* a meaning of ongoing-ness.

In a second instance, in (20), the occurrence figure construes two entities, one (*solvent*) functioning as precipitator and the other (*solid particles*) as precipitated. These are configured with the occurrence (*pull apart*). The concurrent paralinguistic expressions are illustrated in Figure 10.14.

In Figure 10.14, in the faded image on the right and concurrent with *solvent*, a paralinguistic entity is sculpted in two hands. Subsequently on the left and concurrent with the occurrence *pull apart,* two paralinguistic entities, each sculpted in a single hand, move away from each other in contrary directions, depicting an entitied occurrence.

In the spoken language in (20), the entities (*solvent; solid particles)* are technical. The concurrent paralanguage depicts entities in commonsense terms, as things which can be held in the hand/s. The meaning potential is much greater in the

pull apart
solid particles ⟵·················· *solvent*

FIGURE 10.14   Paralinguistic expressions convergent with 'solvent' and 'pull apart our solid particles'

verbal text than in the concurrent paralanguage.[4] However the occurrence, *pull apart,* is construed in commonsense terms. It does not name a specialized activity in the field. Here there is a co-commitment of the commonsense meaning in both language and concurrent paralanguage.

The entities which configure into occurrence figures in text (C) are all construed as technical things (*solution, salt, solvent, solid particles, lattice, water molecules, ion*), and in the concurrent paralanguage the paralinguistic entities are all construed as everyday commonsense things – things held in the hand. The spoken language commits much more ideational meaning of entities than does the concurrent paralanguage. However, the occurrences in text (C) include those which are specialized (*dissolving, uniformly dispersed, being formed*) and others are construed as more everyday and commonsense (*pull apart; surrounds, huddling around; break apart*). In the discourse of text (C), which concerns the building of knowledge of scientific activities, the lecturer moves back and forth between the two. The concurrent paralanguage configures both the entities and the occurrences in everyday terms. In this it functions both convergently and divergently with the spoken language.

Intersemiotic construal of research

As well as the established (and observed) scientific knowledge discussed to this point, a disciplinary field such as chemistry also involves a sub-field of research, which deals with activities of experimentation conducted by scientists. This is exemplified in the language of text (D) where the lecturer gives an account of activities involved in an experiment called 'the hot ice experiment'. For clarity, entities which are pronominally identified are named within round brackets.

> ## text (D)
>
> That experiment there is what we refer to as the <u>hot ice experiment</u>. And [so] what we have there – and what this animation here is showing – is that **we** have **sodium acetate** that **we've <u>added</u>** to **water**, and we've got it to **a saturated point**. **We** then <u>**heated**</u> that **solution** more. [So] **we** <u>**heated**</u> that **solution**, <u>**added**</u> **more and more sodium acetate**, until **it (sodium acetate)**'s <u>**completely dissolved**</u>. **We** can then <u>**let**</u> that **(the solution)** <u>sit</u> at room temperature for ever how long **we** <u>**want**</u>. And it's only after the <u>**addition**</u> of **one single sodium ace-tate crystal** that **it (solution)** <u>**forms**</u> this **crystallization**.

Activities verbally realized though occurrences in (D) are concurrently expressed in paralanguage as motion (as in Figure 10.14). Given the speed of some paralinguistic expressions, it is not always clear whether the rotating hands depict a paralinguistic occurrence figure (of entities and occurrence) or just an occurrence. Unless made clear they are assumed to depict paralinguistic occurrences only. Some such instances are described in Table 10.3 with convergent speech underlined.

Sequencing occurrence figures in language and paralanguage

From a dynamic perspective, in both the subfield of established disciplinary knowledge (text (C)) and that of research (text (D)), we are concerned with how activities are related to one another. In the discourse semantics, occurrence figures are related through either causal or temporal logical connexions (*and then; if… then; so that*) (Hao 2020a; Martin 1992).

Returning briefly to text (C), activities involved in 'dissolving salt' are explained through a series of logically connected occurrence figures. The sequence in (21) is divided into three figures with the connexions linking them shown in small caps.

(21) Those are our water molecules huddling around our different ions
    (IN ORDER) TO break it (lattice) apart.

TABLE 10.3 Convergent expressions of occurrences

Spoken language	Paralanguage
We have sodium acetate that we've <u>added</u> to water	two rotations of the left hand & forearm clockwise away from the body
We've then <u>heated</u> that solution more	five rotations of both hands & forearms one over the other away from the body
<u>added</u> <u>more and more</u> (/kept adding) sodium acetate	two rotations of the right hand & forearm away from the body

AND (THEN) it (lattice) will be uniformly dispersed throughout the solution. The first link is a causal-purposive one – *in order to*. The second is a temporal one – *and then*. The sequence in (21) unfolds by 'implication', meaning that one activity implies what has gone before (Doran and Martin this volume). In the discourse semantics, both temporal and causal connexions can be used interchangeably for construing implication relations (Hao 2020a). This is illustrated in the rephrasing of (21) above as (22) below.

(22) Those are our water molecules huddling around our different ions
    AND THEN (they) break it (lattice) apart.
    SO THAT it (lattice) will be uniformly dispersed throughout the solution.

Text (D) construes the procedure of the *hot ice experiment*. Here, in addition to implication, we also have activities organized through 'expectancy', i.e. one activity expects what potentially follows (Doran and Martin this volume). In example (23), the occurrence figures are related through temporal connexions in small caps.

(23) we heated that solution,
    AND THEN (we) add more and more sodium acetate
    We can THEN let that (the solution) sit at room temperature for however long we want

In contrast to the implication activities, temporal connexions construing expectancy are not interchangeable with causal ones. The rephrasings marked ★ in (24) are not possible.

(24) we heated that solution,
    ★SO THAT (we) add more and more sodium acetate
    ★SO THAT we can let that (the solution) sit at room temperature for however long we want

Language provides resources which can account for a diversity of temporal and causal connexion – resources which allow us to make explicit both expectancy and implication relations in a disciplinary field. In contrast to language, paralanguage cannot differentiate kinds of connexion in a sequence. It can only commit the meaning of connexion as sequence in space/time. This is evident in the relation of the underlined spoken language in text (E) and concurrent paralanguage.

---

text (E)

So the structure of our internal solids is dependent on charge and size – what effect that has on our physical properties – the higher our charges, the greater the electrostatic attraction between positive and negative, which means we're going to need to put more energy in there to break it. Which means we have a stronger bond. <u>So, higher our charge, stronger bond, high melting point.</u>

high melting point .... *stronger bond* .... *higher our charge* ....

FIGURE 10.15 An implication sequence in language concurrently depicted as sequence in space/time only

The underlined summative encapsulation of an implication sequence in text (E), elides all but the grammatical attributes:

> *So, (the) higher our charge (is), (the) stronger bond (is), (the) high(er the) melting point (will be)*

The lecturer's concurrent paralanguage in Figure 10.15 converges with this elided summative encapsulation. It depicts a sequence of three paralinguistic entities located along a horizontal axis in space/time. Potentially these paralinguistic entities, convergent as they are with *higher our charge* etc., might be interpreted as construing qualities, or as semiotic entities realizing a sequence of propositions, *(the) higher our charge (is)* etc.

Activity entities in language and paralanguage

In a final step we return to the discourse semantic system of entities to consider one further type, that of **activity entities** (Hao 2020a, 2020b, this volume). Scientists need to explain and record scientific activities step by step through the sequencing of occurrence figures as in text (E). But they also need to talk about activities in more consolidated terms. They do so by naming activities.

We illustrate the use of activity entities in the lecturer's discourse by referring back to text (C) on 'dissolving salt'. There he explains that dissolving salt in water can make a highly saturated solution, one which reacts to a single added sodium acetate crystal. He illustrates this process in a YouTube video clip which visually displays the reacting process but does not explain it. After the viewing the lecturer consolidates the displayed dynamic field knowledge in the spoken text transcribed in text (F).

text (F)

it's only after the addition of one single sodium acetate crystal that it forms this **crystallization**. [So] we're just seeing that **reaction**. And why we refer to that **reaction** as the **hot ice reaction**, it's because – as we saw in the video – it looks like ice crystals being formed, or feathers, as they so astutely put it in the video. But it's actually an **exothermic reaction**. (…) [So] it's a really useful **reaction**, and it looks kind of cool.

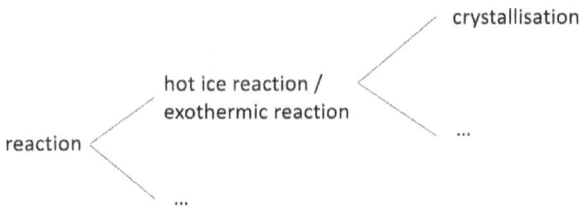

FIGURE 10.16 Types of 'reaction' established in the lecture data

In text (F), the lecturer uses four activity entities to refer to the process seen in the video: *crystallization*, *hot ice reaction*, *exothermic reaction*, and *reaction*. These activity entities are related to one another through classification (Figure 10.16).

The entity *reaction* is a superordinate entity and can be categorized into subtypes. *Exothermic reaction* is one subtype. The entity *hot ice reaction* paraphrases *exothermic reaction* but is not a scientifically defined term. A more delicate subtype of *exothermic reaction* is *crystallization*.

Since activity entities allow for classifying activities, they share similar features with items in field (referred to as itemized activities; Doran and Martin this volume). In classifying activities, activity entities (like technical thing entities) commit a considerable mass of ideational meaning potential.

The inherent relationship between an activity entity and field activities and items makes for interesting interactions with paralanguage. In (25) from text (F), the activity entity in bold, *crystallization*, is concurrently expressed as a paralinguistic entity in Figure 10.17.

(25) it's only after the addition of one single sodium acetate crystal that it forms this **crystallization**.

In text (G), in an explanation of field activities, the lecturer construes two of these as activity entities – *repulsion* and *attraction*.

text (G)

...this is what we know with **ionic solids**. They pack themselves into a certain way in order to maximize **attractions** between **the positive** and **negative** and minimize any **repulsion** between same-charged **ions**.

FIGURE 10.17   The paralinguistic entity concurrent with 'crystallization'

Concurrent with the activity entities of *attraction* and *repulsion* are paralinguistic occurrence figures expressing entities in motion. In our data the paralanguage concurrent with *repulsion* always depicts two sculpted entities (one in each hand) alternating rapidly into and out from the body (Figure 10.18). The paralanguage concurrent with *attraction* is always expressed in the two sculpted entities in rapid horizontal alternating movements from left to right in front of the lecturer's body.

Activity entities in language assign a technical name to field activities. They construe both an activity and an item at the same time.[5] The 'itemized activity' in field can be depicted as either a paralinguistic entity (*crystallization* in Figure 10.17) or a sequence of motion (*repulsion* in Figure 10.18). In either expression this highly technical phenomenon can only have a concrete representation in an everyday, commonsense reality.

FIGURE 10.18 The paralinguistic expression concurrent with of 'repulsion'

Conclusion

Readily observed in the data is the extent to which the semiotic mode of paralanguage is in play throughout the lecture. In convergence with language, meanings are committed in two modes, one supporting the other in meaning-making potential. However, our interest extends to how convergent meanings are being expressed in each mode and the relations which hold between them – the potential offered in the intersemiosis.

Narrowing our focus to concurrence in expressions of ideational meaning (ideational semovergence), two consistent patterns of cooperation and complementarity emerge. The first has to do with the relative commitment of ideational meanings relevant to the established knowledge of the disciplinary field and to building that knowledge in pedagogic encounters. The construal of the field of molecular chemistry in the lecturer's talk is unsurprisingly highly technical overall, although there are shifts into more commonsense from time to time. In the technicality which names entities and occurrences, a heightened mass of ideational meaning is committed. This is complemented in paralanguage in the construal of a commonsense reality, one which commits apparent concrete and tangible entities, entities in motion, and motions across gestural space.

A second perspective on the complementarity is from mode. This points us to textual resources for identifying entities in discourse. Of particular relevance here is the distinction between generic and specific identification. The technicality of the spoken language refers to phenomena in categorical terms which generalize across contexts, while the paralanguage depicts specific instances of phenomena in the immediate shared material reality of the lecture theatre. The intersemiosis of the lecturer's language and paralanguage thus bridges the technical and the everyday, a generalizing perspective and a specific material reality – all in support of building knowledge of the specialized field. It is in both these senses that we discuss the paralanguage as grounding learning.

Finally, we note the rapidly evolving environment in which undergraduate students are introduced to new uncommonsense disciplinary knowledge. A growing reliance on online platforms and social media technologies for pedagogic interaction sees correspondingly reduced access to face-to-face lectures for many students. Justifications for the pace and extent of these changes in pedagogic modes rarely engage with an adequate research base to inform and critique practices, particularly one which takes account of how modes of interaction participate in knowledge building. Our aim has been to contribute insights to inform policies and pedagogic practices which support students to move into new uncommonsense understandings of their disciplinary fields. As we have shown, lecturers' paralanguage in face-to-face teaching/learning offers intuitive yet significant support for students' access to uncommonsense knowledge. We propose that this mode of communication be taken into serious consideration in pedagogic practices.

Notes

1 Comparable observations on embodied meanings are found within the broader gesture literature from other theoretical and disciplinary perspectives. Key sources for other perspectives include McNeill (1992, 2000), Kendon (2004) and Calbris (2011).

2 An additional discourse semantic unit of *quantity* is not discussed in this paper as it was not foregrounded in our selected segments of data.

3 In referring to a congruent construal of meaning in paralanguage we avoid using the term **iconicity** which carries a very different meaning in SFS from that typical in other descriptive accounts of gesture. In the latter it labels a commonsense association of gestures with things they look like in a material world (Wilkin and Holler 2011: 299, McNeill 1992). In SFS, a reference to paralanguage as **iconic** would be making special reference to the way in which everyday commonsense language construes a sense of a world around us as a commonsense ideational reality. In other words, it would necessarily be interpreted through meaning-making choices in language (Ngo et al., 2021).

4 In (21) we infer from the co-text that *solid particles* refer to very many entities, but the commitment of quantity in the paralanguage is limited to two entities, one in each hand.

5 Note this is not grammatical metaphor as there is no possible unpacking at the grammatical level.

References

Calbris, G. (2011) *Elements of Meaning in Gesture*, Philadelphia: John Benjamin.

Cléirigh, C. (2011) *Gestural and Postural Semiosis: A Systemic Functional Linguistics Approach To 'Body Language'*, Unpublished manuscript.

Doran, Y. J. (2018) *The Discourse of Physics: Building Knowledge through Language, Mathematics and Image,* London: Routledge.

Ge, Y.-P., Unsworth, L., Wang, K.-H., and Chang, H.-P. (2018) 'Image design for enhancing science learning', in K.-S. Tang and K. Danielsson (eds) *Global Developments in Literacy Research for Science Education,* Dordrecht: Springer, 237–58.

Halliday, M. A. K. (1978) *Language as Social Semiotic,* London: Arnold.

Halliday, M. A. K. (1994) *An Introduction to Functional Grammar,* London: Arnold.

Halliday, M. A. K. and Greaves, W. S. (2008) *Intonation in the Grammar of English,* London: Equinox.

Halliday, M. A. K. and Martin, J. R. (1993) *Writing Science,* London: Falmer Press.

Halliday, M. A. K. and Matthiessen, C. M. I. M. (2014) *Halliday's Introduction to Functional Grammar,* London: Routledge.

Hao, J. (2018) 'Reconsidering "cause inside the clause" in scientific discourse – from a discourse semantic perspective in systemic functional linguistics', *Text and Talk,* 38(5): 520–50.

Hao, J. (2020a) *Analysing Scientific Discourse from a Systemic Functional Linguistic Perspective,* London: Routledge.

Hao, J. (2020b) 'Nominalisation in scientific English: A tristratal perspective', *Functions of Language,* 27(2): 143–73.

Hao, J. and Hood, S. (2019) 'Valuing science: The role of language and body language in a health science lecture', *Journal of Pragmatics,* 139, 200–15.

He, Y. and van Leeuwen, T. (2020) 'Animation and the remediation of school physics – a social semiotics approach', *Social Semiotics,* 30(5), 665–84.

Hood, S. (2008) 'Summary writing: Implicating meaning in processes of change', *Linguistics and Education* 19, 351–65.

Hood, S. (2010) *Appraising Research,* New York: Palgrave Macmillan.

Hood, S. (2011) 'Body language in face-to-face teaching: A focus on textual and interpersonal metafunctions', in S. Dreyfus, S. Hood and M. Stenglin (eds) *Semiotic Margins,* London: Continuum, 31–52.

Hood, S. (2020) 'Live lectures: The significance of presence in building disciplinary knowledge', in J. R. Martin, K. Maton and Y.J. Doran (eds) *Accessing Academic Discourse,* London: Routledge, 211–35.

Hood, S. and Lander, J. (2016) 'Technologies, modes and pedagogic potential in live versus online lectures' *International Journal of Language Studies* 10(3), 23–42.

Kendon, A. (2004) *Gesture,* Cambridge: Cambridge University Press.

Lemke, J. L. (1990) *Talking Science,* Norwood, NJ: Ablex.

Martin, J. R. (1992) *English Text: System and Structure,* Amsterdam: John Benjamins.

Martin, J. R. (2008) '*Tenderness: Realisation and instantiation in a Botswanan town*', *Odense Working Papers in Language and Communication,* 30–62.

Martin, J. R. (2011) 'Multimodal semiotics: Theoretical challenges', in S. Dreyfus, S. Hood and M. Stenglin (eds), *Semiotic Margins,* London: Continuum, 243–70.

Martin, J. R. (2017) 'Revisiting field: Specialized knowledge in Ancient History and Biology secondary school discourse', *Onomázein Special Issue,* March: 111–48.

Martin, J. R. and Matruglio, E. (2020) 'Revisiting mode: Context in/dependency in Ancient History classroom discourse', in J. R. Martin, K. Maton and Y. J. Doran (eds) *Accessing Academic Discourse,* London: Routledge, 72–95.

Martin, J. R., Rose, D. (2007) *Working with Discourse,* New York: Continuum.

Martin, J. R. and Veel, R. (1998) *Reading Science,* London: Routledge.

Martin, J. R. and Zappavigna, M. (2019) 'Embodied meaning: A systemic functional perspective on paralanguage', *Functional Linguistics,* 6 (1), 1–33.

Martinec, R. (2000) 'Rhythm in multimodal texts', *Leonardo 33* (4), 289–97.

Martinec, R. (2001) 'Interpersonal resources in action', *Semiotica, 135* (1/4), 117–45.

Martinec, R. (2002) 'Rhythmic hierarchy in monologue and dialogue', *Functions of Language,* 9(1), 39–59.

Matthiessen, C. M. I. M. (2007) 'The multimodal page: A systemic functional perspective', in T. Royce and W. Bowcher (eds) *New Directions in the Analysis of Multimodal Discourse,* New Jersey: Erlbaum, 1–62.

McNeill, D. (1992) *Hand and Mind,* Chicago: University of Chicago Press.

McNeill, D. (2000) *Language and Gesture,* Cambridge: Cambridge University Press.

Ngo, T., Hood, S., Martin, J. R., Painter, C., Smith, B. and Zappavigna, M. (2021) *Modelling Paralanguage Using Systemic Functional Semiotics,* London: Bloomsbury.

Painter, C., Martin, J. R. and Unsworth, L. (2013) *Reading Visual Narratives,* London: Equinox.

Saussure, F. (1974) *Course in General Linguistics,* London: Peter Owan.

Wilkin, K. and Holler, J. (2011) 'Speakers' use of "action" and "identity" gestures with definite and indefinite references', in G. Stam and M. Ishino (eds) *Integrating Gesture,* Philadelphia: John Benjamins, 293–307.

Zappavigna, M., Cléirigh, C., Dwyer, P. and Martin, J. R., (2010) 'The coupling of gesture and phonology', in M. Bednarek and J. R. Martin (eds) *New Discourse on Language,* London: Continuum, 237–66.

Zhao, S. (2010) 'Intersemiotic relations as logogenetic patterns: Towards the restoration of the time dimension in hyper-text description', in M. Bednarek and J.R. Martin (eds) *New Discourse on Language,* London: Continuum, 195–218.

11

DOING MATHS

(De)constructing procedures for maths processes

David Rose

Introduction

Reading to Learn (R2L) is both a teaching methodology and a professional learning program that trains teachers in the methodology (Rose 2017). The methodology is designed to enable teachers to support all students in their classes to continually succeed with learning tasks at appropriate levels for their age, grade and curriculum. Central to the methodology is a close analysis of the texts that students are expected to read and write (knowledge genres) and of the classroom activities through which students learn the curriculum (curriculum genres). A primary focus is on guiding students to read curriculum texts, and then use what they have learnt from reading in their writing. Mathematics differs from other curriculum subjects as its written texts are generally brief and ancillary to the major curriculum genre of the discipline – which is in fact the intermodal modelling of maths processes by teachers, followed by individual student problem solving using the modelled process.

While R2L offers strategies for teaching the written maths genres of definitions, explanations and 'word problems', the primary focus of the professional learning program and this chapter is on the curriculum genre modelling maths processes. The term 'maths process' is commonly used by teachers for mathematical genres that Doran (2018) describes technically as *quantifications*, which find a numerical result for a maths problem, and *derivations*, which derive relations among technical symbols. We will use the name *maths process modelling genre* for the curriculum genre that models these mathematical genres.

Following the introduction, the chapter describes this curriculum genre as it has been re-designed in the R2L methodology. The re-designed process modelling genre has three stages before students are expected to solve problems for practice and assessment. The first is a teacher demonstration of the process using a detailed lesson plan. The second is a series of guided practice cycles with increasing handover of control to students. The third is a joint teacher/student construction of the procedural steps for the process. Each of these stages is described by analyzing extracts from a secondary school maths lesson. Lessons are analyzed from three perspectives: the structuring of pedagogic activities, teacher/learner interactions, and the modalities through which meanings are exchanged in the discourse. This approach is known as pedagogic register analysis (Rose 2014, 2018).

The primary social concerns of R2L are with the inequity of educational outcomes and with enabling teachers to narrow the achievement gap between their students. This inequity is particularly apparent in maths outcomes. At the level of systems, there seems to have been little or no growth in maths outcomes in recent generations (Leigh and Ryan 2008, Meeks, Kemp and Stephenson 2014). Concern is regularly expressed in the media about the diminishing rate of candidates for science-based degrees, all of which depend on expertise in maths. At the classroom level, Table 11.1 presents the averaged outcomes of maths topic tests in one Sydney secondary maths class. While the top group is consistently successful, the middle group barely passes, and the botton group consistently fails. Maths teachers in the R2L professional program report that this scale of difference in achievement levels is typical. The proportion of students achieving success in maths may vary between classes and suburbs, but the majority of students are still not successful enough to qualify for maths dependent disciplines. For the bottom group to achieve passable scores would require an average growth rate of over 60%. For the middle group to reach the top group average would require over 30% growth rate. This was the challenge for designing effective maths curriculum genres in R2L.

Teachers in the R2L professional program regularly express three specific concerns about their students' maths learning. One is with the 'maths word problems' that are supposed to contextualize students' calculation tasks in their everyday experience. Text 1 is an example from the trigonometry section of a Year 9 maths textbook (Barnes *et al.* 2015: 188).

Text 1: Maths word problem

> *A rescue helicopter spots a missing surfer drifting out to sea on his damaged board. The helicopter descends vertically to a height of* 19m *above sea level and drops down an emergency rope, which the surfer grips. Due to the wind the rope swings at an angle of 27° to the vertical, as shown in the diagram. What is the length of the rope?*

A major concern of teachers is the effect on weaker students of all these irrelevant words, which distract students from the abstract data and operations they need to solve the problems. It appears that this pervasive practice may actually have the opposite effect of its purported function, to facilitate maths learning. As students are assessed on their success with problem solving, the emotional effect on weaker students is to create and maintain learner identities as 'no good at maths'. At the level of schools' socioeconomic function to reproduce educational hierarchies (the assessment 'bell curve'), it may be surmised that this too is an actual social function of maths word problems. In response, R2L has designed a simple analysis of maths

TABLE 11.1 Average % scores in topic tests in Year 8 maths class

Student	%
Samim	90
Syed	89
Rafah	85
Usaama	75
Timothy	70
Blake	61
Daniel	61
Top group average	**76%**
James	59
Dylan	57
Sameer	56
Luke	55
Ryan	54
Jack	53
Marcus	50
Zack	50
Middle group average	**54%**
Daniel	47
Airrison	41
Monroe	39
Matthew	36
Cory	33
Steven	28
Thomas	27
Brendan	25
Bottom group average	**34%**

word problems and a procedure for applying it in the classroom that enables students to disentangle the data, solution and operations from their wordings. Briefly, the three-step method is to first identify the data presented literally by mathematical items in the question, e.g.

- height of 19m
- angle of 27° to the vertical

The second step is to infer the required solution, encoded in lexical relations among wordings. In this case, this is the length of the hypotenuse, which must be inferred from lexical relations between *height* 19m *above sea level* <–> *rope at angle* 27° *to the vertical* <–> *length of rope*. The third step is to choose the maths processes required to solve this question, in this case, calculating the hypotenuse from sine of 27° with opposite side of 19m.

Another concern of teachers is with students' inability to comprehend, remember and apply technical terminology in maths. One source of this problem is the complex technicality of mathematical texts, such as the definitions and explanations that traditionally introduce each topic in maths textbooks. A typical example is Text 2, from the Year 9 trigonometry curriculum.

Text 2: Technical definition

In any right angled triangle, for any angle:

The sine of the angle	=	the length of the opposite side
		the length of the hypotenuse.
The cosine of the angle	=	the length of the adjacent side
		the length of the hypotenuse.
The tangent of the angle	=	the length of the opposite side
		the length of the adjacent side.

Where students are weak readers, one strategy of maths teachers is avoid reading such highly technical texts in textbooks, and instead present definitions and explanations orally, in the course of modelling relevant maths processes on the board. This in fact seems to be the standard practice in the majority of Australian classrooms. A recent trend in textbooks is to delete many technical texts, and instead present technical terms in lists. A problem that remains with either approach is that written questions in problem-solving activities are equally technical, so that weaker students still struggle to read them.

In response, R2L provides teachers with a method known as Detailed Reading and Rewriting. In these activities, teachers guide students to identify each meaning in a short passage of text and discuss their meanings in depth. Detailed Reading enables all students to read a text with comprehension, no matter what their independent reading abilities. The information from the text is typically then written on the board as notes, and the class then rewrites the notes as a new text. When maths teachers regularly apply these strategies to key texts, students rapidly learn to independently read mathematical technicality. Detailed Reading and Rewriting are designed curriculum genres, described in Rose and Martin (2012).

The maths process modelling genre

A third teacher concern is the inability of many students to independently solve problems, following teacher modelling of the relevant process. This is perhaps the most common curriculum genre in maths classrooms the world over. In a typical practice, the teacher demonstrates each step of the process by writing a worked example on the board, explaining what they are doing, and engaging the class by asking for appropriate information at each step. Close observation shows this practice to be a highly complex mix of modalities and interactions between teachers and students. This pedagogic complexity is additional to the process itself, which may involve many steps and choice points, along with technical terms and knowledge that students must have mastered in previous lessons and school stages.

For this reason, some teachers choose to model difficult processes with more than one worked example and involve students as much as possible, before they are expected to solve problems individually. Nevertheless, all teachers participating in R2L professional learning report that a minority of students in their classes can generally solve problems accurately following the modelling. Other students have more difficulty with problem solving, while others cannot do the tasks without individual support from the teacher. Although teachers spend most time supporting these students during problem solving activities, they continue to struggle with maths. The pedagogic function of problem solving activities is to practise applying maths processes. Hence successful students receive most benefit from the practice, and weak students achieve least benefit. Furthermore, the practice function is conflated with evaluation, so that weaker students experience continual failure – the emotional impact of which further reduces their learning capacities and learner identities.

The source of these disparities in students' capacity to apply modelled processes to problem solving is not hard to see. Students are more likely to succeed with problem solving if they are able to follow teachers' explanations in each step of a process, understand the technical terms they are using, recognize the relations between the verbal explanation and the symbolic text and diagrams that are being written on the board, and remember each step. If students miss any of this complex web of elements, they will find problem solving more difficult.

The R2L program addresses this problem by redesigning the maths process modelling genre, to provide more support for all students to learn processes effectively. The goal of this redesigned genre is for students to record the steps in a maths process as a written procedure, which they can then apply to problem solving. To this end, teachers plan how they will model the process in close detail, including the approximate wordings they will use to explain each step, the questions they will ask students, and the worked example they will write on the board. This detailed planning impels teachers to carefully consider how they present maths processes, and the wordings they use to do so. Table 11.2 shows such a lesson plan for the trigonometry lesson analyzed in this chapter.

The first column names eight steps that this teacher considers necessary for solving trigonometry questions at this school grade. These steps will be named orally as the process is demonstrated in the lesson. The second column includes one or more questions with which students' participation may be engaged. These questions are designed to ensure an appropriate answer, derived from information written on the board or shared knowledge from previous lessons. The third column shows how the teacher will elaborate these responses as a series of notes, diagram and equations on the board.

Teachers then use such procedural plans to model the maths process. This modelling is similar to their standard practice, demonstrating and explaining each step of a process, except that it has been carefully planned. The process is then modelled

TABLE 11.2 Lesson plan for R2L maths process genre

Steps	Questions	Example
1 Read the question.	What does it say?	In a triangle LMN, angle M = 90°. Side LN is 9.2m and side LM is 8.2m. Find the angle L to the nearest degree.
2 Make a note of the important information in the question.	What is the first important information?	The triangle is a right angled triangle
	What is the next important information?	Side LM is 8.2 Side LN is 9.2 Right angle at M
3 Draw a labelled diagram with this information.	What are the angles called?	
	What are the lengths of the sides?	
4 Label the angle to be found as Θ	Which angle is Θ?	

5 Write down the three trigonometric ratios related to the right angled triangle.	What is the first trigonometric ratio?	$\text{Sin } \Theta = \dfrac{\text{opposite side}}{\text{hypotenuse}} = \dfrac{MN}{9.2}$
	What is the next ratio? What is the next ratio?	$\text{Cos } \Theta = \dfrac{\text{adjacent side}}{\text{hypotenuse}} = \dfrac{8.2}{9.2}$
		$\text{Tan } \Theta = \dfrac{\text{opposite side}}{\text{adjacent side}} = \dfrac{MN}{8.2}$
6 Find the right ratio to solve the question.	Which ratio gives both sides?	$\text{Cos } \Theta$
7 Use this ratio to solve the question.	What are the steps to simplify the ratio?	$\dfrac{\cos \Theta}{1} = \dfrac{8.2}{9.2}$ $9.2 \times \text{Cos } \Theta = 8.2$
	What is the value of Θ?	$\Theta = \text{Cos}^{-1} \dfrac{8.2}{9.2}$
8 Use a calculator to calculate the angle.	What is the answer?	$\Theta = \text{Shift Cos } \dfrac{8.2}{9.2}$ $\Theta = 27^0$

again with a different example, but this time the teacher asks students to name each step, and students take turns to write the worked example on the board, with the help of the teacher and class. This guided practice is then repeated, typically once for secondary students and up to five or six times for younger students, with increasing handover of control to students in each iteration. Finally, students take turns to record each step of the process, as a written procedure on the board, with the help of teacher and class, and students write the procedure in their maths books for later reference. Results to date include 100% improvements within three months for failing students, and an average of 20% improvements within two months in controlled research projects (Lövstedt and Rose 2015, Table 11.10 below).

In terms of genre and register theory, each of these iterations – demonstration, guided practices and joint construction – are stages in the R2L maths process curriculum genre. Each step in the lesson plan is a phase, involving a series of activities, within each genre stage. Stages are structural units at the stratum of genre, that realize one or another genre. Phases are structural units at the stratum of register, realizing features in field, tenor and mode systems (Martin and Rose, 2007, Rose 2006, 2020).

The methodology is freely available for teachers and researchers in a set of videoed demonstration lessons and resources, produced by this author for the NSW Education Standards Authority (NESA 2018). Extracts from these videoed lessons are analyzed here to describe the stages of the curriculum genre.

Pedagogic register analysis

As previewed earlier, the lesson analyses have three functional strands. They describe the structuring of pedagogic activities, that are negotiated in pedagogic relations between teachers and learners, and presented through pedagogic modalities of speaking, writing, signing, drawing, viewing, gesturing and other somatic activity. This is an analysis at the level of register. In broader terms, pedagogic activities are the field, pedagogic relations are the tenor, and pedagogic modalities are the mode of a pedagogic register.

The institutional function of a pedagogic register is to exchange knowledge and values which constitute a curriculum register. Learners construe, or abduce, curriculum knowledge through the pedagogic register, which also enacts curricular values. Curricular values include learners' degrees of access to knowledge through pedagogic modalities, the authority they derive from this access, the affirmation they receive, their inclusion in the learning conversation, and the learning autonomy they acquire. Each learner's abduction of curriculum knowledge is conditioned by their positioning in these value hierarchies, in relation to their peers.[1] While curriculum knowledge is often an explicit focus of lessons, curricular values tend to be enacted tacitly.

A curriculum genre is an unfolding configuration of features in pedagogic register and curriculum register. In Bernstein's terms (1990: 62), the function of curriculum genres is 'cultural reproduction or transformation'. The purpose of pedagogic register analysis is to describe how these social processes unfold at the level of individual learners and classes; to get at how learning actually occurs, by closely observing pedagogic exchanges using the research tools of systemic functional theory (Rose 2014, 2018).

In the lesson analyzed here, the curriculum register includes knowledge of the procedure to solve trigonometry questions. Its values are carefully oriented towards equality of access, authority, affirmation and inclusion. The goal is equal autonomy in the problem-solving practice that follows. In terms of instantiation theory (Martin 2008), the procedure presented in the lesson re-commits a mathematical *quantification* genre (Doran 2018) as a series of verbally named steps. Steps 1–3 are concerned with the Situation stage of this genre (the question and its data), steps 4–7 with the Substitution stage (substituting numerical values in symbolic formulae), and step 8 with the Numerical Result.

In terms of individuation, curricular values may unequally allocate resources to learners to abduce and apply the procedure, hence their readings of the procedure may be divergent. The goal of this lesson is to allocate access, authority, affirmation and inclusion more equally, to facilitate convergent readings of the procedure, and hence more equal success with autonomous problem solving.

The global structuring of pedagogic activity comprises a hierarchy of segments, from whole lessons, to lesson activities, to learning cycles at the micro-level of teacher/learner interactions. Units at each of these levels may be iterated in series. The lesson analyzed here is one in a series within the curriculum topic of trigonometry. Each stage of the lesson comprises a series of lesson activities named as steps in the lesson plan (Table 11.2). Each of these lesson activities comprises a series of learning cycles that are concerned with a specific micro-activity, such as writing each trigonometric ratio in step 5. This hierarchy is illustrated for the trigonometry lesson in Figure 11.1. In this curriculum genre, the central stage of *Guided practice* may also be complexed into series, indicated in the figure by superscript[n].

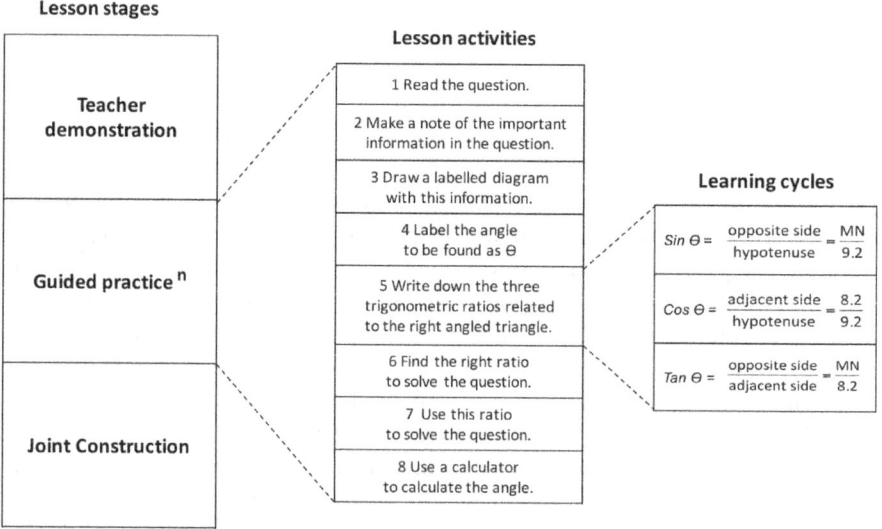

FIGURE 11.1 Hierarchy of lessons stages, lesson activities and learning cycles

FIGURE 11.2 Nuclear and marginal phases of learning cycles

The structure of units at each of these levels is centred on a learning task, through which knowledge is abduced by learners. Tasks may involve an active display by learners, to identify meanings in a text or propose from their own knowledge. Or tasks may be receptive, to receive meanings verbally, or perceive them visually. Learning tasks are typically focused (specified) and then evaluated by a teacher. In addition, teachers may first prepare learners to succeed with the task, and the knowledge they construe through the task may then be elaborated. These five structural elements are termed Prepare, Focus, Task, Evaluate and Elaborate phases (Martin and Rose 2007, Rose and Martin 2012). Their orbital structuring as nuclear and marginal phases is diagrammed in Figure 11.2. Analyses may also name the pedagogic matter that each phase is concerned with. This enables the analyst to identify relations between the matter of the learning Task, and Prepare, Focus and Elaborate phases.

Pedagogic relations include the roles of teachers and learners in interactions, and the pedagogic acts that they negotiate. Pedagogic roles are termed *interacts* in the analyses, as the term *role* is used within discourse semantics, for generalized speaker roles in exchange structures. Interacts specify the pedagogic functions of exchange roles at the connotative stratum of register. For teachers, potential interacts include presenting knowledge, evaluating learners and directing the activity. Learners may display knowledge and both learners and teachers may solicit acts from each other. Acts include various pedagogic behaviours and acts of consciousness such as attention, perception, knowledge, reasoning, engagement and anticipation.

Pedagogic modalities are the sources of meanings in the learning discourse, and the means of sourcing them into the discourse. Sources of meanings include the environment, spoken knowledge of teachers and learners, and records such as written texts, graphic images and video recordings. Each of these source types has a set of options for sourcing them into the discourse. For example, written texts may be read aloud, but meanings in images must be recast as wordings. Meanings in the environment and in records may also be indicated by naming or pointing to them, by words or gestures. Spoken teacher/learner knowledge may be individual or shared, and may be presented or elicited by teachers, and recalled or inferred by learners. In addition, meanings may be recorded during the lesson, and then become sources. They may be recorded by writing verbal or symbolic texts, or annotating a

FIGURE 11.3 Multimodal teacher demonstration

text or image, or by drawing pictures or diagrams. Many of these options in modalities are visible in Figure 11.3, a still from the videoed lesson (NESA 2018), in which the teacher points at the hypotenuse in the diagram, while saying the trigonometric ratio, *sine theta is equal to opposite divided by hypotenuse*. Values in pedagogic activities, relations and modalities, that are applied in the analyses here, are set out as tables in the Appendix to this chapter.

Teacher demonstration stage

Table 11.3 is an extract from step 2 in the teacher demonstration, in which key information from the question is written as notes on the board. At this point the teacher has read out the question which is written on the board (step 1). This extract illustrates how students may be engaged in a maths demonstration by asking for their perception, knowledge or reasoning. Where possible, realizations of interacts are italicized in the transcript.

The extract includes two learning cycles, marked by horizontal lines in the table. The first cycle begins with the teacher preparing the step, by directing the activity to come, *you write down all the important points from the question*. The task is for students to identify items in the question, with the property *all the important points*. This requires them to reason about which items are important. To this end, the focus phase invites students to reason, *Who can tell me what's the first important point*.

This focus phase includes the 'hands-up' classroom ritual, marked by dashed lines. The command *Put your hand up* insists that students display for teacher evaluation, the hand-up gesture invites evaluation, and one student is then permitted to display. The curricular values of this ritual are inclusion and affirmation of

TABLE 11.3 Teacher demonstration, Step 2: writing the ratios

			phase	*interact*
1	T	Step 2 says *you write down* all the important points from the question.	prepare step	direct activity
	T	*Who can tell me what's* the first important point?	focus property	invite reasoning
		Put your hand up		insist display
	Ss	[hands up]		invite evaluation
	T	Yes, S1		permit display
	S1	Right angle triangle	identify property	display reasoning
	T	First it's a right angle triangle.	affirm	repeat
		[writes *It is a right angled triangle*]	elaborate step	
		It is a right angle triangle.		
		First important point.		
2	T	Next one?	focus item	inquire reasoning
	Ss	[hands up]		invite evaluation
	T	[points to S2]		permit display
	S2	Angle M is ninety degrees	identify item	display reasoning
	T	Angle M is ninety degrees.	affirm	repeat
		[writes $\angle M = 90°$]	elaborate step	

students' authority. By commanding the whole class, the teacher here is countering a widespread problem where only a few students in a class are willing to invite evaluation.

Here, Student 1 displays reasoning by identifying the item, *Right angle triangle*. The teacher affirms by repeating this display, and then elaborates by writing the item on the board, reading it back, and restating its value as the *first important point*. This cycle combines curriculum goals for knowledge and values. The student's response instantiates the procedural step with a particular example. In so doing, he models the reasoning task for the whole class to see. By repeating the display, the teacher both affirms the student and reinforces the item. It is further reinforced by writing it and reading it again, which models the activity, and is again valued as *first important point,* which reinforces the reasoning task in this procedural step.

Cycle 2 reiterates the same pattern more briefly. The focus simply inquires reasoning, *Next one?* Student 2 is permitted to display reasoning and is affirmed by repeating the display, *Angle M is ninety degrees.* This time the elaboration recasts this verbal sentence by writing a mathematical equation, a symbolic text, $\angle M = 90°$. These symbols will be used to label the diagram in step 3 of the trigonometry process.

Table 11.4 is an extract from step 4 of the demonstration stage, in which trigonometric ratios are written using values from the diagram. This monologic

TABLE 11.4 Teacher demonstration, Step 4: writing the ratios

		phase	*sourcing*
1	OK, Step 4. [**writes** *Step* 4]	prepare step	new teacher write note

2	Since this is a trigonometry question
	   we're going to
	 write down all the trig ratios related to
	   the question.
	 [**points** to diagram]
		classify text
		
		 new teacher
		
		 point diagram	

3	We'll write down sine theta, cos theta and tan theta. [**lists** on fingers]	restate lesson symbolize text	
4	So, **looking from** the sine of **theta**, [**writes** *Sin θ =*] sine theta is equal to [**points** to diagram]	focus expression perceive expression receive expression perceive part	locate diagram write expression recast expression point diagram
	opposite [**points** to opposite line]	receive expression perceive part	restate lesson point diagram
	divided by hypotenuse. [**points** to hypotenuse]	receive expression perceive part	restate lesson point diagram
5	Therefore, MN [**writes** *Sin θ = MN*] divided by	receive expression perceive expression receive expression	recast diagram write expression recast expression
	hypotenuse [**points** to hypotenuse] which is 9.2. [**writes** *Sin θ = MN/9.2*]	receive expression perceive part receive expression perceive expression	restate lesson point diagram recast diagram write expression

demonstration involves a series of rapid shifts in modality, from speaking to writing words and mathematical expressions, to indicating the diagram, to speaking and writing again.

Learning tasks in this cycle are receptive, switching back and forth from perceiving written expressions and parts of the diagram to receiving the teacher's spoken knowledge. In the table, each sentence in the monologue is numbered, and dashed lines mark couplings of speaking and writing or gesturing.[2] Where possible, sources and sourcing are bolded in the transcript.

In sentences 1–3, the teacher prepares the step, first by stating *Step* 4, and writing it as a note on the board, then by classifying the question, *a trigonometry question*, and presenting new teacher knowledge about the step, *to write down all the trig ratios related to the question*, and pointing to the diagram. He then specifies these ratios as *sine theta, cos theta and tan theta*, which restates shared knowledge from prior lessons.

In Sentence 4, the teacher reviews the elements of the first ratio, *the sine of theta,* that is shared knowledge from prior lessons, while pointing to the relevant elements in the diagram. The focus locates the ratio in the diagram, *looking from the sine of theta.* The students' first tasks are then to **perceive** this as a symbolic expression *Sin θ =,* to **receive** it rephrased in wordings, *sine theta is equal to,* and **perceive** the relevant part of the diagram as the teacher points to it. Their next tasks are to **receive** the technical term *opposite,* while **perceiving** this part of the diagram as the teacher points. The next tasks are to **receive** the expression *divided by hypotenuse,* and **perceive** this part of the diagram as teacher points.

In Sentence 5, the teacher specifies these elements of the ratio with values from the diagram. The first task is to **receive** the expression *Therefore MN,* which recasts labels from the diagram. (*M* and *N* are labelled angles in the diagram, so *MN* is the side of the triangle between these angles, opposite the angle *θ.*) The next tasks are to **perceive** this expression as the teacher writes it, including the underline *MN,* that symbolizes division, and **receive** this operator as words, *divided by.* The final tasks are to **receive** the term *hypotenuse,* and to **perceive** this part of the diagram as the teacher points to it, then to **receive** its value, *which is* 9.2, and **perceive** the whole ratio as the teacher finishes writing it, *Sin θ = MN/9.2.*

The intricate complexity of this intermodal activity illustrates a major source of difficulties that many students experience with maths processes. It should be pointed out that the teacher is an expert practitioner. He carefully reviews and builds on knowledge that has been accumulated in previous lessons and uses it to contextualize the worked example with which he illustrates the new process. The rapid modality shifting is designed to facilitate this contextualization and is undoubtedly effective for many learners. It maximizes their access to the curriculum knowledge, by recommitting the procedure and its content through multiple modalities.

Nevertheless, if a student has a weak command of any of the spoken, written, symbolic and visual elements, or relations between them, they may struggle to maintain the close attention needed to follow the teacher's modelling. Consequently, teachers must often re-explain the process to such students during the individual problem solving activity. This pervasive practice attempts to **repair** the ineffectiveness of standard curriculum genres, after students have failed to acquire the knowledge they need, and simultaneously evaluates them as failing. The R2L methodology reverses this normal but inequitable practice, by ensuring that all students are **prepared** to succeed with learning tasks before they are evaluated. In the re-designed maths process genre, this is accomplished with a series of guided practices of the process, and a joint construction of the procedure. It does not matter if any students did not get a perfect grasp of the procedure from the teacher demonstration, as it is now repeated with increasing handover of authority, towards autonomy for all.

First guided practice

Table 11.5 is an extract from Step 2 of the first guided practice in this lesson. There are now few prepare phases in learning cycles, as the tasks have been prepared in the

TABLE 11.5 First guided practice, Step 2: labelling the diagram

			phase	sourcing	interact
1	T	*What is* the **next step?** [hands up] S1	focus step	remind lesson	inquire knowledge / invite evaluation / permit display
	S1	Draw the diagram	propose step	recall lesson	display knowledge
	T	Draw the diagram. OK. Let's see how you can draw the diagram.	affirm / elaborate step		repeat / invite display
	S1	[comes to board, **draws** triangle]		draw diagram	display knowledge
2	T	*Who can tell him* where to put **B?** [hands up] S2	focus place	read text	invite reasoning / invite evaluation / permit display
	S2	Bottom left	propose place	locate diagram	display reasoning
	T	Bottom left.	affirm		repeat
	S1	[**labels** angle B]	elaborate step	label diagram	display knowledge
3	T	Where would **C** go? [hands up] S3	focus place	read text	invite reasoning / invite evaluation / permit display
	S3	Bottom right	propose place	locate diagram	display reasoning
	T	Bottom right.	affirm		repeat
	S1	[**labels** angle C]	elaborate step	label diagram	display knowledge

FIGURE 11.4 Labelling the diagram

prior demonstration. Instead of the teacher demonstrating, authority is handed to the students, who now take turns to come to the board to scribe notes, expressions, diagrams and labels. In Cycle 1, the focus reminds students of knowledge presented in the demonstration, *What is the next step?* Student 1 proposes this step by recalling the demonstration, *Draw the diagram.* The teacher affirms by repeating, and invites the student to display drawing skills, *Let's see how you can draw the diagram.* Drawing the right angle triangle recasts this wording as a graphic image, and recalls the demonstration, further displaying the student's knowledge.

In Cycles 2 and 3, student 1 starts labelling the diagram from notes on the board (Figure 11.4), guided by the class. The teacher's first focus is a place in the diagram, *where to put B.* Student 2 locates this place in the diagram, *bottom left,* by reasoning from the note *right angle at B,* to the right angle in the diagram. Cycle 3 repeats this pattern with angle *C,* as do the following cycles until the diagram is fully labelled with values given in the question.

Table 11.6 analyses step 4, to show how this complex intermodal activity is negotiated as students take more authority. The first focus invites students to display their recall of the step, *writing the ratios.* Student 1 first recalls the formula from the demonstration and prior lessons, *Cos theta equal to adjacent over the hypotenuse,* and then recasts it with values from the diagram, *which is BC over 30.* The teacher affirms by directing him to '*come up*' to the board, where he recasts his answer as a written equation, $Cos\theta = BC/30$.

Cycle 2 follows a similar pattern, but student 2 proposes only the type of ratio, *sine.* Consequently, the teacher prepares cycle 3 by restating this move, and then focuses the required expression, *Sine theta over what?* Student 2 then proposes these values from the diagram, *22 over 30.* Again, the teacher affirms by directing him to '*go ahead*' to the board, where he recasts the values as a written equation, $Sin\theta = 22/30$. Cycle 4 is now even more abbreviated. The focus simply names the next

TABLE 11.6 Writing the ratios in first guided practice

			phase	*sourcing*	*interact*
1	T	*Who can come up and write* **all the ratios** *at the same time?*	focus step	remind lesson	invite display
	Ss	[hands up]			invite evaluation
	T	S1			permit display
	S1	Cos theta equal to adjacent over the hypotenuse	propose formula	recall lesson	display knowledge
		which is BC over 30.		recast diagram	display reasoning
	T	Come up.	affirm		direct display
	S1	[comes to board, **writes** $Cos\Theta = BC/30$]	elaborate equation	write expression	display knowledge
2	T	*Who can* do the **next ratio?**	focus expression	remind lesson	invite display
	Ss	[hands up]			invite evaluation
	T	S2			permit display
	S2	Sine	propose expression	recall lesson	display knowledge
	T	OK	affirm		approve
3	T	S2 *says* sine ratio.	prepare expression	restate move	repeat display
		Sine theta over *what?*	focus expression		inquire reasoning
	S2	That's 22 over 30	propose expression	recast diagram	display reasoning
	T	Go ahead.	affirm		direct display
	S2	[comes to board, **writes** $Sin\Theta = 22/30$]	elaborate equation	write expression	display knowledge
4	T	Tan theta?	focus expression	remind lesson	inquire reasoning
	Ss	[hands up]			invite evaluation
	T	S3			permit display
	S3	Tan theta which is 22 over BC	propose equation	recall lesson	display reasoning
	T	Tan theta is 22 over BC.	affirm	recast diagram	repeat
	S3	[comes to board, **writes** $Tan\Theta = 22/BC$]	elaborate equation	write expression	display knowledge

ratio, *Tan theta?* Student 3 proposes the values 22 *over BC* from the diagram, the teacher repeats, and the student comes to the board without being directed, to write the equation, $Tan\theta = 22/BC$.

In this manner, the intermodal activity of recalling the formula for each ratio, identifying values in the diagram, and writing these as an equation, is negotiated interactively between teacher and students. Students are invited to display their authority and are consistently affirmed.

In this first guided practice, the teacher chooses students who are most likely to recall the formulae and reason accurately about the relevant values in the diagram, following his demonstration. However, the whole class experiences this previously modelled activity reinstantiated as an interactive process involving their peers. In this stage around half the students in the class have an opportunity to propose expressions or write on the board. The second guided practice now includes students who are typically less successful at maths.

Second guided practice

Table 11.7 analyses writing the ratios in the second guided practice, in which the teacher uses more interpersonal scaffolding to engage weaker students, along with the support provided by repeating the activity. Cycle 1 is similar to cycles in the first guided practice. Student 1 recalls the first ratio, adding values from the diagram, *Cos theta equal to* 17 *over PR*, and is directed to write it on the board.

However, in Cycle 2, the teacher first prepares by directing students to reason about the next ratio, warning them to be ready, *I'm going to ask you about the next ratio.* He then names student 2, a weaker student, without the hands-up ritual. S2 correctly proposes the ratio with values from the diagram. To affirm S2, the teacher hands authority to the whole class, by asking them to concur with the answer, *He said sine theta is equal to PQ over 21. Is he right?* This is a strategy for inclusion in the learning community. S2 then correctly writes the equation on the board, $Sin\theta = PQ/21$cm. Following all this scaffolding, Cycle 3 abbreviates the interaction. The teacher names student 3, who proposes merely the values from the diagram, *PQ over* 17cm, and then writes the whole equation, $Tan\theta = PQ/17$cm.

The power of repeated guided practice, in which all students participate actively in constructing the maths process, is illustrated in Table 11.8, the solution phase of the process. In cycle 1, multiple students propose the next step, *solve the equation.* The whole class then solicits the weakest student 1 to become the scribe for this step. The teacher checks and asks the student whether he concurs, *Really? You want to go there?* S1 at first demurs by shaking his head, but the other students encourage him by applauding, and he comes to the board and writes the first expression of the solution, $Cos\,\theta =$, displaying his knowledge of the correct ratio to solve the problem. In cycle 2, the teacher asks for the values to '*end up*' the equation. Student 2 proposes 17cm *over* 21, from the notes on the board. The teacher repeats this to S1, who writes it as an expression, completing the equation, $Cos\,\theta = 17/21$.

TABLE 11.7 Second guided practice, Step 4: writing the ratios

			phase	source
1	T	Who can remember the first ratio?	focus expression	remind lesson
	Ss	[hands up]		
	T	S1		
	S1	Cos theta / equal to 17 over PR	propose equation	recall lesson
	T	OK.		recast diagram
		Then write down that here.	affirm	
	S1	[comes to board, **writes** $Cos\ \theta = 17\text{cm}/21\text{cm}$]	elaborate equation	write expression
2	T	Think about the next ratio while he's writing. / I'm going to ask you about the next ratio. / S2, what's the next ratio?	prepare step	remind lesson
	S2	Sine theta	focus expression	recall lesson
		equals PQ over 21	propose equation	
	T	He said sine theta is equal to PQ over 21. / Is he right?	affirm	recast diagram
	SS	Yes		
	T	Go and write on the board.	elaborate equation	
	S2	[comes to board, **writes** $Sin\ \theta = PQ/21\text{cm}$]		write expression
3	T	What's **the next** ratio? S3	focus expression	remind lesson
	S3	PQ over 17 cm	propose expression	
	T	Can you put that on the board there?	affirm	recast diagram
	S4	[comes to board, **writes** $Tan\ \theta = PQ/17\text{cm}$]	elaborate equation	write expression

TABLE 11.8 Second guided practice, Step 8: solving the problem

		phase	interact
1	T	*What is* **the next** step?	inquire knowledge
	Ss	Solve the equation	display knowledge
	Ss	[name and point to S1 (weak student)]	invite display
	T	Really?	check
		You want to go there?	inquire accordance
	S1	[shakes head, smiling]	demur
	Ss	[applaud]	insist display
	S1	[comes to board, **writes** *Cos θ =*]	elaborate expression	display knowledge
	T	OK	affirm	approve
2	T	*What is he* going to end up? S2	focus expression	inquire reasoning
	S2	17cm over 21	propose expression	display reasoning
	T	OK	affirm	approve
		He said 17 over 21.		direct display
	S1	[**writes** *Cos θ* = 17/21]	elaborate equation	display knowledge
	T	OK		approve
3	T	*What is* **the next** step?	focus step	inquire reasoning
	S3	Cos and then to the power of minus 1	propose expression	display reasoning
	T	Anything on the left hand side?	focus expression	inquire knowledge
	S3	Yes, and then theta is equal to	propose expression	display knowledge
	T	OK	affirm	approve
		He's saying *put down* theta is equal to.		direct display
	S1	[**writes** *θ* =]	elaborate expression	display knowledge
5	T	What's he saying?	focus expression	inquire reception
	S4	Cos to the power of minus 1	propose expression	display reception

(Continued)

TABLE 11.8 (Continued)

		phase	*interact*
T	Cos to the power of minus 1.	affirm	repeat
S1	[**writes** $\theta = cos^{-1}$ (17/21)]	elaborate equation	display reasoning
T	Is he right?	affirm	inquire accordance
Ss	Yes		concur
T	OK		approve
6 T	*What is* **the next** step? S5	focus step	inquire knowledge
S5	Using a calculator to calculate the equation	propose step	display knowledge
T	Using calculator to calculate the equation.	affirm	repeat
S5	[uses calculator]	elaborate step	
	The answer is 35 degrees, 57 minutes		display perception
T	35 degrees, 57 minutes.		direct display
S1	[**writes** $\theta = 35°\ 57'$]		display knowledge
T	Give him a round of applause.	affirm	praise
Ss	[applaud]		concur

FIGURE 11.5   Writing the solution

Cycles 3–5 negotiate the next equation in the solution. Student 3 proposes the expression *Cos and then to the power of minus* 1. The teacher asks for the left-hand term in the equation, and S3 proposes *theta is equal to*. The teacher repeats this to S1, who writes the correct expression θ =. In Cycle 5, the teacher asks for the right-hand value to be restated, *What's he saying?*[3] Student 4 restates *Cos to the power of minus* 1, which the teacher repeats to S1, who completes the equation as $\theta = \cos^{-1}$ (17/21). This displays mathematical reasoning, as $\cos^{-1}$ is multiplied by brackets enclosing 17/21. To affirm S1, the teacher then hands authority to the whole class, *Is he right?*

Cycle 6 negotiates the last step in the solution. Student 5 uses his calculator to calculate the value of $\cos^{-1}$ (17/21) and reads the answer as 35 *degrees*, 57 *minutes*. The teacher repeats this to S1, who recasts it correctly in symbols, $\theta = 35° \; 57'$ (Figure 11.5). By means of this collaborative scaffolding, a reluctant student is actively included in modelling the trigonometry process, displays his own authority and is affirmed by his peers and teacher. His achievement is finally affirmed by a round of applause from the class.

Joint construction of procedure

The final stage in the R2L maths process genre is a joint construction of the procedure for the process. Each step in the procedure distils students' experience of the series of activities that they have just observed and participated in. As in the guided practice stages, students take turns to scribe the steps on the board, with the support of their peers and the teacher (Figure 11.6).

Table 11.9 illustrates the negotiation of the joint construction. Each step is negotiated in a single learning cycle, in which the teacher focuses the step and elaborates

FIGURE 11.6 Scribing the procedure

a student's response, and the student writes the step on the board. Cycle 1 begins with preparing by directing the activity, while reminding of the previous lesson stage, *We are going to review the steps.* The focus inquires the first step, student 1 recalls *You read the question,* and he is directed to the board, where he writes the step. Cycle 2 follows the same pattern, minus the preparation phase. Here the teacher expands student 2's answer, but the student writes his own version. Cycle 3 again follows this pattern. Here the teacher again expands student 3's answer, and the student writes the expanded version.

The joint construction is completed in four more steps that follow these variations. Students copy the procedure into their maths books as it is being written, to refer to during individual problem solving. Text 3 shows the completed procedure.

Text 3: Jointly constructed procedure

Trigonometry
Subtopic: Right angle triangle
    Step 1: Read the question
    Step 2: Write the important information
    Step 3: Draw a labelled diagram
    Step 4: Label the angle θ
    Step 5: Write down the three trigonometric ratios
    Step 6: Use process of elimination on 3 trig ratios
    Step 7: Use the right ratio to find θ

TABLE 11.9 Joint construction of procedure steps

			phase	sourcing	interact
1	T	*We are going to* **review the steps** *that you're going to use to solve the right angle triangle.*	prepare activity	remind lesson	direct activity
		Who can tell me, what is **the first thing you do**?	focus step	remind lesson	inquire knowledge
	S1	You read the question	propose step	recall lesson	display knowledge
	T	Very good.	affirm		praise
		Come and put it on the board.	elaborate text		direct display
	S1	[comes to board, **writes** *Step 1: Read the question*]		write text	display knowledge
	T	OK	affirm		approve
2	T	*What is* **the next step you take** in order to solve this problem?	focus step	remind lesson	inquire knowledge
	S2	Write the important information	propose step	recall lesson	display knowledge
	T	Write down the important information from the question.	elaborate text	rephrase move	repeat
		Can you put it on the board?			direct display
	S2	[comes to board, **writes** *Step 2: Write the important information*]		write text	display knowledge
3	T	*What is* the third thing you do?	focus step	remind lesson	inquire knowledge
	S3	Draw a diagram	propose step	recall lesson	display knowledge
	T	Draw a diagram.	affirm		
		Draw a labelled diagram.	elaborate text	rephrase move	repeat
	S3	[comes to board, **writes** *Step 3: Draw a labelled diagram*]		write text	display knowledge

Note that the wordings in Text 3 are not identical with the steps in the teacher's lesson plan (Table 11.1), as Text 3 is jointly constructed with contributions from students as well as teacher. The joint construction with students scribing is a critical activity for students to take control of the procedure, and remember how to reproduce it in their problem solving. Although each step in the procedure is only a synopsis of the activity it stands for, students' participation in the guided activities and recasting them as the procedure ensures that this knowledge becomes their own.

Conclusion

Mathematics is the archetypal intellectual discipline, the exercise of pure reason applying theory. But its practices do not occur just in disembodied minds; they are spoken in words and written in symbols, in order to be shared, extended and accumulated over generations. Moreover, learning mathematics is an emotional experience, and the sources of emotion are almost always found in our relations with others. In this regard, the function of affirmation in pedagogic practice cannot be overstated. Bernstein tells us that "evaluation condenses the meaning of the whole [pedagogic] device" (2000: 50). Where the social function of this institution is to create and maintain inequality, it uses unequal evaluation to create unequal inclusion and learner autonomy. Hence the curriculum registers of mathematics are more than mathematical knowledge such as trigonometry formulae and quantification genres; equally important are curricular values of access, authority, affirmation, inclusion and autonomy.

Pedagogic relations are not simply a knowledge transfer of transmission and acquisition. Rather learners abduce knowledge and values from the totality of an unfolding pedagogic register, and this abduction inevitably varies from learner to learner. The degrees to which learners are able to abduce elements of culturally accumulated knowledge such as trigonometry are directly related to the curricular values they experience. By middle secondary school, successful learners have internalized continual affirmation. They can learn autonomously, not only because they possess the requisite field knowledge (authority) and reading skills (access), but because they are self-affirming.

Such students are able to attend to the rapid modality switching in teacher demonstrations, illustrated in Table 11.4 above, and reproduce the activity in their independent problem solving. This targeted repetitive practice with continual success automatizes their procedural knowledge of the process so that it can be applied with ease to complex problems that require meta-reasoning about procedures. The neurological basis of such procedural learning and reasoning is described by Edelman (1992), who explains how perceptual categorization is necessarily charged by emotional values from the brain's limbic system. More abstract concepts are formed by the brain's perceptions of its own categories, again charged with emotional values. Automatization results from repeated, emotionally charged perception and concept formation, that becomes the background for further perceptual categorization and concept formation.

Here is another concern of teachers, and a criticism of the R2L curriculum genre described above. Teachers report that struggling students are unable to interpret the solutions required by complex maths problems, even when they know the processes involved. The constructivist position, that currently dominates western teacher education, holds that learners must first understand the maths 'concept', from which they can find their own way to solve problems. From these positions, the R2L genre appears to be constraining rather than enabling, as it teaches a predefined set of steps that cannot help students' creative problem solving. What is invisible to these positions is the automatized procedural knowledge possessed by successful students, that enables them to focus attention on meta-procedures to interpret complex problems.

The R2L curriculum genre is designed to provide access to this procedural knowledge for all students in 'mixed-ability' classes (i.e. all classes). Its informing principles include the model of learning cycles, in which learners are prepared for learning tasks so that they are continually successful and affirmed, and ready for the elaborations that follow. Second is the principle of repetition with increasing handover of authority from teacher to learners. Third is the need to ensure that every student in a class is equally included and affirmed in the learning activity. Fourth is the intermodal strategy of recasting spoken knowledge as written texts, and vice versa, in both teacher planning and jointly constructed procedures.

The goal is to maximize the autonomy of all students to practise individual problem solving successfully, and so benefit from its pedagogic functions. To this end, R2L principles are also applied to problem solving activities. Teachers select representative problems and guide the class to solve them jointly – identifying data, solution and operations – before individual practice with the remaining problems. This strategy further minimizes potential struggles for weaker students and maximizes success and automatization for all. For teachers, the close analysis of learning tasks, and of their own teaching practice, becomes an automatic part of their practice. It can then be applied to analyzing and explicitly teaching the meta-procedures involved in complex problem solving, rather than mystifying them as exotic cognitive processes. For students, the kinds of results that can be expected from consistent application of the R2L maths process genre are illustrated in Table 11.10. This table shows the growth in average scores in maths topic tests over two school terms, for the same Year 8 students as Table 11.1. Significantly, the average growth rates for each cohort are close to the necessary growth rates predicted by the achievement gaps in Table 11.1.

Appendix: Pedagogic Register Analysis

Table 11.11–11.15 set out the values in pedagogic activities, modalities and relations that are applied in analyses above. See Rose (2018) for further discussion.

TABLE 11.10 Growth in topic test scores using R2L maths process genre

Student	Pre-R2L average scores	After 2 terms of R2L maths	Percentage growth
Samim	90	93	3
Syed	89	95	7
Rafah	85	90	6
Usaama	75	82	9
Timothy	70	80	14
Blake	61	83	36
Daniel	61	68	11
Average growth of top group			**12%**
James	59	68	15
Dylan	57	55	0
Sameer	56	74	32
Luke	55	73	32
Ryan	54	87	61
Jack	53	89	68
Marcus	50	54	8
Zack	50	63	26
Average growth of middle group			**30%**
Daniel	47	51	8
Airrison	41	53	29
Monroe	39	67	46
Matthew	36	48	33
Cory	33	66	100
Steven	28	55	96
Thomas	27	39	44
Brendan	25	67	128
Average growth of bottom group			**60%**

TABLE 11.11 Learning cycle phases

Nuclear Phase	focus	guided	explicit criteria	
		unguided	criteria implicit	
	task	manual	...	
		semiotic	displaying	identify propose	in text from knowledge
			receptive	receive perceive	verbally visually
	evaluate	affirm reject (if not affirmed)		
Marginal Phase	prepare	with explicit criteria before focus		
	elaborate	monologic dialogic	teacher knowledge negotiated with learners	

TABLE 11.12 Record sources

Record sources			record modality		
			verbal	visual	
	record type	graphic record	verbal text symbolic text	picture diagram	
		recording	audio recording	video recording	
	record access	individual			
		shared	display copy		
Record sourcing	sourcing mode	restate	repeat	e.g. read aloud	
			diverge	summarize rephrase recast	
		indicate	pointing	verbal gesture	locate point
			describing	verbal	compare class/part
				gesture	imitate symbolize
	sourcing language	same as record			
		other language			

TABLE 11.13 Spoken sources

Spoken sources	individual knowledge	teacher knowledge	
		learner knowledge	
	shared knowledge	prior lesson prior move	
Spoken Sourcing	teacher speaking	present	new restate
		elicit	remind enquire
	learner speaking	recall infer	

TABLE 11.14 Recording

Write			writing mode		
			wordings	symbols	
	writing type	write text	constructed text notes	equation expression	
		annotate record			
Draw	picture diagram mark record				

TABLE 11.15 Acts and interacts

Interacts	teacher roles	instructing	presenting knowledge	model				
				impart				
				check (before evaluating)				
			evaluating	evaluate				
			learners		affirm	repeat		
						approve		
						praise		
					reject	implicit	qualify	
							ignore	
						explicit	negate	
							admonish	
		directing	direct					
			suggest					
			permit					
	learner roles	display	(for evaluation)					
		accord	concur⁴					
			demur					
	teacher/learner roles	invite						
		inquire						
		insist						
Acts	behavioural acts	learners	behaviour					
			display					
			accordance					
		teacher	evaluation					
		teacher/learners	activity					
	conscious acts	perceptive	attention					
			perception					
			reception					
		cognitive	knowledge					
			choice					
			reasoning					
			conception					
		affective	attitude					
			engagement					
			anticipation					

Notes

1 Hierarchies in curricular values expand Bernstein's 'hierachy of success and failure' (2000: 11), in terms of curriculum genres, where access:pedagogic modalities :: affirmation and inclusion:pedagogic relations :: authority:curriculum knowledge :: autonomy:control over curriculum genres.
2 In terms of discourse semantics, written sentences in lesson transcripts here are moves in exchanges (Martin and Rose 2007, Rose 2018). Each role of teacher and learners in the exchange may be complexes of moves, or single items such as *OK*. Each exchange move has a register level value in pedagogic relations. Learning cycle phases may be realized by one or more exchange moves. Pedagogic modalities tend to switch from move to move, but can also switch rapidly within moves, or be constant across several moves. Linguistic labelling of exchange moves is not included in the analyses here, to reduce the complexity of the presentation and save space in tables.
3 Doran (2018) describes the 'left-hand' and 'right-hand' of equations as Theme and Articulation of mathematical statements.
4 Learners have options to concur or demur from teachers' or learners' acts, whereas teachers have the institutional authority to evaluate learner acts by affirming or rejecting (Bernstein 1990, Rose 2018).

References

Barnes, M., J. Bradley, L. Elms and R. Cahn (2015) *Maths Quest 9 for the Australian Curriculum*, 2nd edition, Brisbane: Jacaranda Press

Bernstein, B. (1990) *The Structuring of Pedagogic Discourse*, London: Routledge.

Doran, Y. J. (2018) *The Discourse of Physics: Building knowledge through language, Mathematics and Image*, London: Routledge.

Edelman, G. (1992) *Bright Air, Brilliant Fire: On the matter of the mind*, London: Harper Collins.

Leigh, A. and Ryan, C. (2008) '*How has School Productivity Changed in Australia?*' Canberra: Australian National University http://econrsss.anu.edu.au/~aleigh/.

Lövstedt, A-C and Rose, D. (2015) *Reading to Learn Maths: A teacher professional development project in Stockholm*, https://www.readingtolearn.com.au/wp-content/uploads/2016/01/Reading-to-Learn-Maths-A-teacher-professional-development-project-in-Stockholm.pdf.

Martin, J. R. (2008) 'Tenderness: realization and instantiation in a Botswanan town', Odense Working Papers in Language and Communication, 30–62.

Meeks, L., Kemp, C. and Stephenson, J. (2014) 'Standards in literacy and numeracy: Contributing factors' *Australian Journal of Teacher Education (Online)*, 39(7), 106.

NESA (2018) *Planning for Success in Secondary Mathematics*, Sydney: NSW Educational Standards Authority, http://educationstandards.nesa.edu.au/wps/portal/nesa/k-10/learning-areas/mathematics/planning-for-success-in-secondary-mathematics.

Martin, J.R. and Rose, D. (2007) *Working with Discourse: Meaning beyond the Clause*, London: Continuum.

Rose, D. (2006) 'Reading genre: a new wave of analysis', *Linguistics and the Human Sciences* 2(2), 185–204.

Rose, D. Analysing pedagogic discourse: an approach from genre and register. *Functional Linguist.* **1**, 11 (2014). https://doi.org/10.1186/s40554-014-0011-4.

Rose, D. (2017) 'Languages of schooling: Embedding literacy learning with genre-based pedagogy' *European Journal of Applied Linguistics*, 5(2), 1–31.

Rose, D. (2018) 'Pedagogic register analysis: Mapping choices in teaching and learning' *Functional Linguistics*, 5(3), 1–33.

Rose, D. (2020) 'The baboon and the bee: Exploring register patterns across languages', in J. R. Martin, Y. J. Doran and G. Figuedero (eds) *Systemic Functional Language Description*, London: Routledge, 273–306.

Rose, D. and Martin, J. R. (2012) *Learning to Write, Reading to Learn: Genre, knowledge and pedagogy in the Sydney School*, London: Equinox.

INDEX

Note: concepts from Legitimation Code Theory or Systemic Functional Linguistics are indicated as LCT or SFL